NAFA GUIDE TO AIR FILTRATION

National
Air
Filtration
Association

Fourth Edition, 2007

National Air Filtration Association
PO Box 68639
Virginia Beach, VA 23471

Publication Information

The *NAFA Guide to Air Filtration, Fourth Edition* is published by the National Air Filtration Association (NAFA). NAFA is a nonprofit, industry-supported trade association whose members include air filter sales and service companies, manufacturers and distributors of air filtration products, as well as those associated with the air filtration industry.

Additional copies of *NAFA Guide to Air Filtration, Fourth Edition* are available from NAFA Headquarters at www.nafahq.org, or by writing to NAFA, PO Box 68639, Virginia Beach, VA 23471.

First Edition, 1993

Second Edition, 1996

Third Edition, 2001

Fourth Edition, 2007

ISBN No: 978-0-938423-22-5

NAFA encourages architects, engineers and end-users to contact NAFA-member companies for air filter information, products and services. NAFA Members subscribe to the NAFA Code of Ethics, study and pass a thorough international examination to achieve accreditation (Certified Air Filter Specialist – CAFS, and/or NAFA Certified Technician – NCT), and receive continuing education and training on "best practice" applications and emerging technologies in the air filtration industry.

A complete list of NAFA Members can be found at the NAFA website at www.nafahq.org.

NAFA Membership

Membership in NAFA is open to individuals involved in the sales, service, manufacture, and distribution of air filter products and service. In addition, NAFA has other categories of involvement for those associated in any way with clean indoor air. For further information, log onto the NAFA website at www.nafahq.org.

Foreword

Air filtration is one of the most important components in an HVAC System. Properly engineered and correctly applied air filtration provides a facility and its occupants with protection from particulate- phase and molecular-phase contaminants that are both a nuisance and can cause illness and even death. It also provides protection to expensive mechanical equipment, interior furniture and fixtures, industrial processes and products that require high levels of cleanliness in manufacture.

This 4th Edition of the NAFA Guide to Air Filtration is the most comprehensive revisions since the original edition was printed in 1993. The Technical Committee of the National Air Filtration Association considered a number of factors in this revision that included current state of the industry, intent of the manufacturers and scope of air filter distributors along with the user community's access to internet-based information. It is written in both Inch-Pound (IP) and Standard International (SI) systems of measurement (see Appendix One for Systems, Formulas and Conversions).

The American Society of Heating, Refrigerating and Air Conditioning Engineers (ASHRAE) and the International Organization for Standardization (ISO) continues to research and clarify air filter testing guidelines and standards and indoor air quality standards for buildings.
This edition references the latest information from these and other standard and rule making authorities and organizations.

There are many more pictures, diagrams and charts showing newer technologies and larger varieties of air filtration products, many designed to meet Leadership in Energy and Environmental Design (LEED) criteria. Sustainability is also emphasized with additions to Chapter 13 – Owning and Operating Costs - on Life Cycle Costing of Air Filtration. Two new chapters have been added (to reflect newer technologies (Chapter 14 - UVC and PCO) and industrial applications (Chapter 15 - Industrial Contaminant Air Filtration Control Systems).

The 4th Edition is required study for those individuals wishing to complete the NAFA Certified Air Filter Specialist (CAFS) accreditation. The international test available in both English and Spanish is administered periodically throughout the year in order to further the educational quality in connection with sales and distribution of air filter products and services.

For additional information or comments and questions about the information in this book, or for a listing of NAFA Members in your area, please go to the NAFA web site at www.nafahq.org or contact NAFA Headquarters located in Virginia Beach, VA at 757-313-7400.

Acknowledgments

Special recognition and acknowledgement for publication of the 4th Edition must be given to the following groups and individuals:

All Members of the National Air Filtration Association

Contributors to Previous Editions:

Robert Avery
David Books, CAFS
William Down, CAFS
Gerald Festian, CAFS
Robert Hanson, CAFS
Skiter Kowalski, CAFS
Chis Muller
Thomas Ottney
Bill Rasmussen, CAFS, NCT
Tom Riddell, CAFS
Dave Sickels, CAFS
George Spottswood, CAFS
Gail Waite, CAFS
Rex Wilhoite
Robert Williams, CAFS

The 2007 NAFA Technical Task Group, NGAF Rewrite Committee:

Chair
Jeff Gentry, CAFS
Ahlstrom Air Media
West Chester, OH

Vice Chair
Harry Allen, CAFS
Lifetime Industries, Inc.
Memphis, TN

Members:
Todd Brizendine
AAF, International
Louisville, KY

Harry Elinsky, Jr., CAFS
Filtech, Inc.
West Homestead, PA

Joe Fly, Jr. , CAFS
Joe Fly Company, Inc.
Dallas, TX

Dave Goss, CAFS
Filtration Solutions, Inc.
Waukesha, WI

Jim Hedback, CAFS
Hedback Quality Air
Indianapolis, IN

Todd Hensley
Menardi Filters

Rupert Langston, CAFS
Air Flow Technology, Inc.
Kenosha, WI

Paula Levasseur, CAFS
Cameron Great Lakes, Inc.
Portland, OR

Jack Manns
Hollingsworth & Vose
East Walpole, MA

Dave Matela, CAFS
Kimberly Clark Corporation
Roswell, GA

Ron Mattson, CAFS
DP Systems
Addison, IL

Steve Nicholas, CAFS, NCT
Air Industries, Inc.
North Andover, MA

Domenick Orlando, CAFS
Dave Downing & Associates
Phoenix, AZ

Rick Peckham
Clarcor Air Filtration
Louisville, KY

Leslye Sandberg, CAFS
Permatron
Elk Grove Village, IL

Al Vatine, CAFS
LMS Technologies
Bloomington, MN

Chris Zaker, CAFS, NCT
Glasfloss Industries
Dallas, TX

A special thanks to the following people who contributed their scientific, technical skills and knowledge to specific chapters:

David Brande
Biosafety Consultant

Jason Downing, CAFS
Dave Downing & Associates

Karin Foarde
Research Triangle Institute

David Hurd, CAFS
PlymoVent North America

Wayne Ketner, CAFS
Dave Downing & Associates

Bill Patterson, CAFS, NCT
Dave Downing & Associates

Amy Ragland
Honeywell Inc.

Jeffrey Siegel, Ph.D.
Department of Architectural and Environmental Engineering
University of Texas at Austin

Dean Tompkins
University of Wisconsin

Dan Young
IAQ Energy

A special thanks to NAFA Association Partners and other organizations who cooperated by allowing parts of their standards, guidelines and references to be used in this text:

- American Conference of Governmental Industrial Hygienists (ACGIH)
- American Society of Heating, Refrigerating and Air Conditioning Engineers (ASHRAE)
- Institute of Environmental Science and Technology (IEST)
- International Organization For Standardization (ISO)
- National Sanitation Foundation (NSF)
- Underwriter's Laboratories, Inc. (UL)

NAFA Headquarters Staff:

Alan C. Veeck, CAFS
Executive Director

Terry Driscoll
Executive Assistant / Graphic Designer

Joe Abel
Computer Graphic Designer

Table of Contents

Chapter 6: Air Cleaners 6.1

Chapter 7: HVAC Filter Testing 7.1

Addendum 7.1 Addendum 7.1.1

Chapter 8: HEPA and ULPA Filter Testing 8.1

Chapter 12: Indoor Air Quality 12.1

Chapter 13: Owning & Operating Costs 13.1

Addendum 13.1 Addendum 13.1.1

Chapter 14: Ultraviolet Germicidal Irradiation, Photocatalytic Oxidation 14.1

Chapter 15: Industrial Contaminant Air Filtration Control Systems 15.1

1

The Importance of Air Filtration

Introduction

Comfort air conditioning is described as "the process of treating air to simultaneously control its temperature, humidity, cleanliness, and distribution to meet the comfort requirements of the occupants of the conditioned space.[1]" Air conditioning for uses other than comfort is classified as "industrial air conditioning." The four requirements of temperature, humidity, cleanliness, and distribution control apply equally to industrial air conditioning.

The Air We Breathe

Air is a mixture of gases composed of approximately 21% oxygen, 78% nitrogen, 1% argon and carbon dioxide, and traces of other gases (Figure 1.1). The air we breathe also includes particulate material and gases generated by nature, by man, and by industrial processes. We are concerned with the particulate matter and gases that influence our health or comfort, that damage the spaces we occupy, or that affect the products or components we are manufacturing.

Although great strides have been made in cleaning up the environment, the air we breathe is not very clean. Pages could be used to list all the sources and nature of the materials that pollute it. Once, outdoor air was considered the main source of particulates found indoors. However, the oil crisis in the 1970's, started an effort to save energy resulting in less and less outdoor air being used in building air conditioning systems. It was found that this resulted in a buildup of contaminants within occupied spaces and created indoor air quality problems. These contaminants come from building materials and finishes, building contents such as furniture and fixtures, and from people, processes and materials within the building. Many of these are nuisance particulate and molecular contaminants such as cooking odors, but others are identified as causing discomfort and even illness for some, if not all, of the building occupants. Regardless of its source, an airborne contaminant can be either an aerosol or a gas.

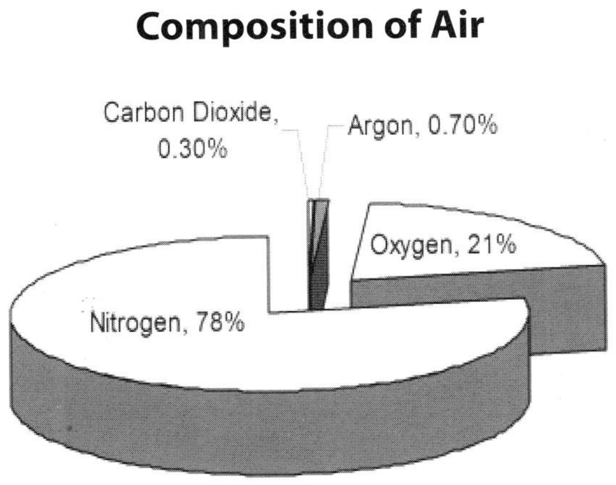

Figure 1.1. The Air We Breathe

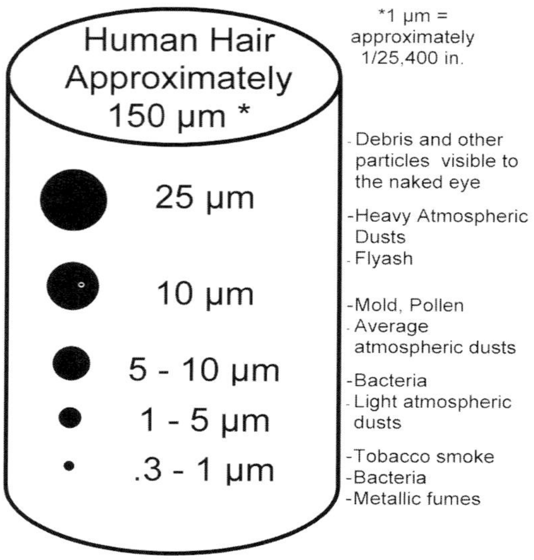

Figure 1.2. Size in µm of Certain Small Particles

Aerosols

Aerosols cover a wide range of sizes as shown in Figure 1.6. Particle size is the most important property of an aerosol because the smaller the particles, the more stable the aerosol and hence, the greater the difficulty in separating the particles from the gas phase in which they are suspended.

An aerosol is a suspension of solid or liquid particles in the air.

The size of an aerosol is usually measured in micrometers which is also abbreviated "µm". A µm is one millionth of a meter or approximately 1/25,400 inch. Figure 1.2 helps visualize the size of a µm by relating this dimension to the size of a human hair and other objects and Figure 1.3 is actual particles magnified on filter fibers.

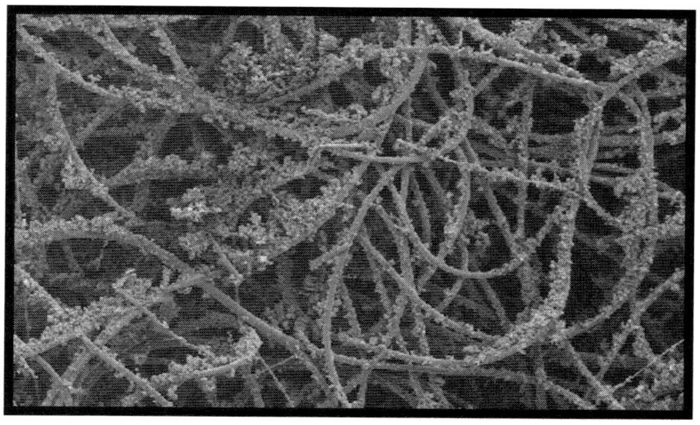

Photo Courtesy Kimberly Clark Corporation

Figure 1.3. Close-up of Particles on Filter Fibers

Solid Particles

Dusts are solid aerosols generated from the reduction of larger solid materials. As examples, a jackhammer creates dust while drilling holes in a rock, and a volcanic eruption discharges tons of fine lava dust into the air.

The Environmental Protection Agency (EPA) classifies ambient atmospheric particles into two primary modes, coarse and fine. Coarse particles, which include those larger than 3 µm, are generally formed by mechanical breakup of solids, primarily natural and chemically inert. Fine particles are those formed from chemical reactions or condensing gases. They have a maximum size of 1-3 µm and are more chemically complex than coarse particles (Figure 1.4)[1].

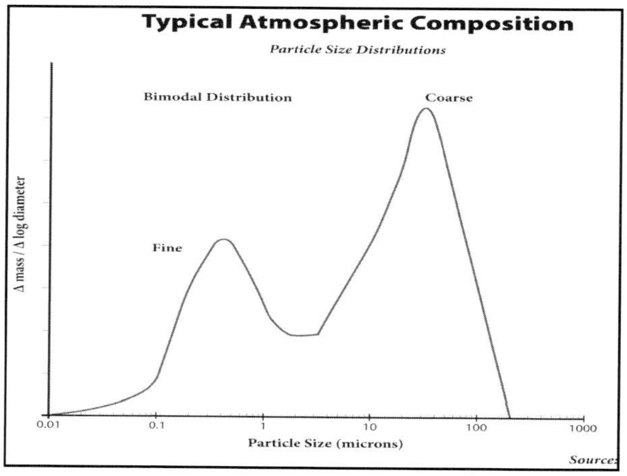

Source U.S. EPA

Figure 1.4. Typical Atmospheric Composition

Although, larger dust particles settle rapidly, smaller dust particles tend to stay suspended in the air or to settle very slowly. According to the ASHRAE Handbook[2] airborne dust particles less than 0.1 µm behave like gases and have no rate of fall but are affected by Brownian Motion. Those in the range of 0.1 to 1.0 µm have negligible settling velocities, while those in the range of 1.0 to 10 µm have constant and appreciable settling rates but are kept in suspension by air currents. Particles larger than 10 µm will normally settle out of the atmosphere. See Figure 1.5.

Particle Settling Velocities	
Particle Sizes µm	**Settling Velocity**
100.00	59.2 feet per minute
50.0	14.8 feet per minute
10.0	7.1 inches per minute
5.0	2.5 inches per minute
1.0	5.1 inches per hour
.05	1.4 inches per hour
0.1	1.13 inches per day
<0.1	negligible

Figure 1.5. Shows average settling velocities of particles in still air.

Bioaerosols are airborne biological materials including fungi, bacteria (and their spores) and viruses.

Fumes are solid aerosols formed by the condensation of vapors of solid materials. Arc welding fumes are a typical example. The heat of the electric arc is enough to vaporize some of the rod and its coating. When they cool, they form welding fumes. Very small fume particles have a tendency to agglomerate forming larger particles.

Liquid Particles

Fogs are liquid aerosols formed by the condensation of water vapor in the air.

Mists are liquid aerosols formed by the atomization of liquids.

Compound Particles

Smokes are solid and/or liquid aerosols formed by the incomplete combustion of organic substances.

Smog is a compound mixture of particulates and droplets seen as a haze caused by a sunlight-induced photochemical reaction. Smog most often results over a populated area when a temperature inversion restricts normal removal of contaminants.

Environmental Tobacco Smoke (ETS) is a compound mass of poly-dispersed smaller sized particles, fumes and vapors created from burning tobacco products.

Measuring Particulate Contamination

The concentration of different size particles in the atmosphere is usually measured by weight or count. Methods for testing filters are based on these two techniques.

Weight concentration can be determined by drawing a measured amount of air through a pre-weighed filter paper target. The paper can then be reweighed to determine its weight increase. This increase represents the weight of the dust in the quantity of air which was drawn through the filter paper. Knowing the total weight of dust and the volume of air sampled, one can calculate the weight concentration of dust for a unit volume of air.

Counting and sizing of dust particles in the atmosphere were once tedious tasks. Initially, air samples were drawn through impingers in which the airborne dust was captured in a measured amount of water or some other liquid. Representative samples of the liquid containing the dust in suspension were then examined under a microscope and the different size particles counted. The introduction of membrane filters simplified this procedure. Air samples were drawn through membrane filter papers which collected the dust on their surface. The membranes could be treated with liquids which made them transparent and allowed the dust particles captured on them to be counted and sized under a microscope. The use of membrane filters is still a recognized method for counting and sizing particles.

The commercial development of the optical particle counter (OPC) has simplified the measurement of concentration and size of particles in the atmosphere. The original devices, using tungsten light for illumination, were limited to measuring particle sizes down to 0.3 µm. Newer units, using lasers as a source

Figure 1.6. Comparative Particle Size Chart

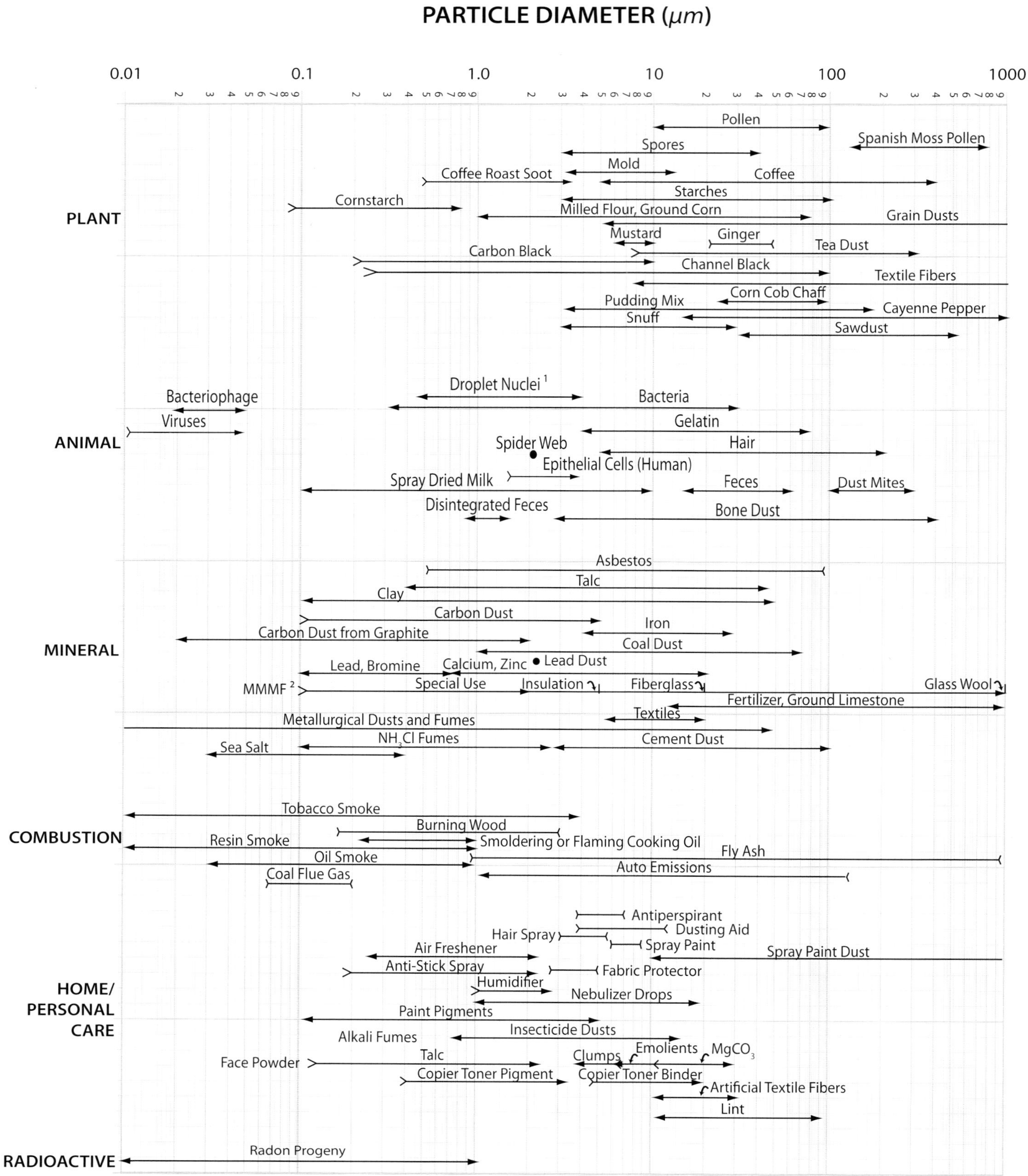

[1] Liquid droplets containing bacteria etc. sneezed etc.
[2] Man-made mineral fibers

Owens, Ensor, Sparks

of light, have the capability of counting particles smaller than 0.1 µm, and Condensation Nuclei Counter have the ability to measure in the nanometer range (0.001 µm). Determination of dust concentration on a count basis has become extremely important, especially in precisely controlled environments such as cleanrooms. Cleanroom cleanliness is defined by its class. For instance, according to ISO 14644-1, a Class 3 cleanroom can have no more than 102, 0.3 µm particles per cubic meter of air.[3] For more detail, see Chapter 9: "*Controlled Environments*."

When one measures the staining effect of a dust, an area measurement is involved. Even though every solid dust particle has three dimensions, staining only involves two of them; the length and width of the particle as it lies on the surface it is staining. The relative concentration of the staining fraction of dust in the atmosphere can be measured by the discoloration of a filter paper through which a known amount of air has been drawn. No positive numbers result from such a test, but the relative staining characteristics of environment "A" can be compared to that of environment "B."

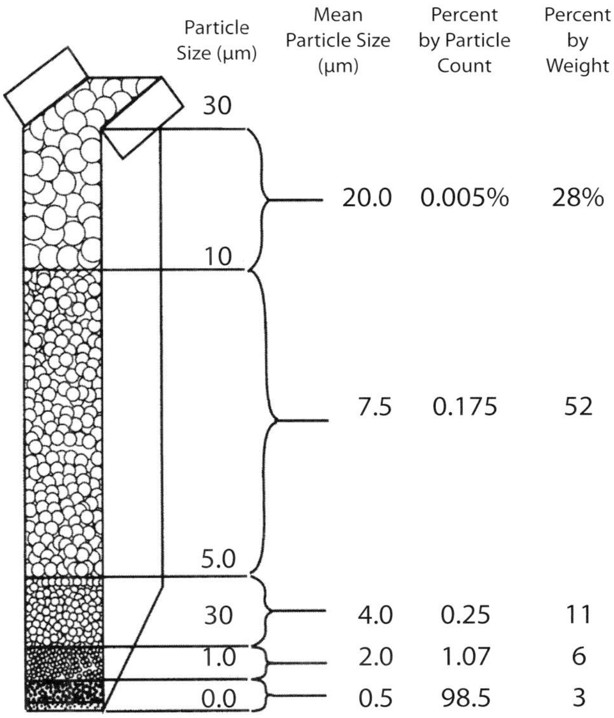

Figure 1.7. Particle Size Distribution in the Atmosphere[4]

Effect of Measuring Method

Work done under the sponsorship of ASHRAE and private organizations such as filter manufacturers showed that the percentage distribution of particle sizes in the atmosphere depended significantly on the method of measurement. A variation of one illustration[4] used to show the relationship between particle size distribution and measurement is shown in Figure 1.7. For instance, particles smaller than 1.0 µm comprise only 3% by weight of all atmospheric dust particles, but make up 98.5% by count. On the other hand, atmospheric dust particles in the 5.0 to 10 µm size range comprise 52% by weight, but only 0.175% on a count basis. Note that almost all the particles in the outdoor atmosphere are less than 1 µm in diameter.

Airborne Gases

A gas is any material that has the tendency to expand indefinitely and which completely and uniformly fills the container it occupies. Gases exist as molecules. Examples are oxygen, nitrogen and carbon dioxide.

Vapors are gases formed by the evaporation of materials that are normally liquid or solid.

In addition to the normal gases in the atmosphere, the air we breathe contains a variety of gases in different mixtures and concentrations depending on our location. Some of these gases have odors which are pleasant such as the fragrance from roses, while others are disagreeable such as hydrogen sulfide which smells like rotten eggs. Many gases are corrosive or toxic. Fortunately, in most instances, the human odor threshold (concentration at which the odor of a gas is recognized) is much lower than the level at which a gas becomes irritating, corrosive, or toxic. There are, however, some gases which are odorless and do not alert a person to their toxicity. Carbon monoxide is the best known in this category.

Source of Gases

Gases are formed in chemical processes where they are deliberately prepared for other uses or where they may be the unwanted by-product of a process. In the first case, leakage in process equipment may cause the escape of gas to the atmosphere. In the second case, the unwanted by-product may be deliberately vented to the atmosphere.

Gases also occur as a result of biological processes. The digestion of sewage sludge by bacteria may produce large quantities of odorless methane, but it also produces smaller amounts of corrosive and highly odorous organic sulfur compounds. The production of certain antibiotics is accompanied by highly odorous by-products.

As mentioned previously, one more recently identified source of gases in the form of volatile organic compounds (VOCs) has been from plastics and other materials used in the manufacture of floor coverings, furniture fabrics, and other decorative materials. This is discussed in more detail in Chapter 11: "*Airborne Molecular Contaminants.*"

Measurement of Concentration of Gases

The measurement of the concentration of a gas in the atmosphere depends on the specific gas. The concentration of some gases can be measured easily, while that of others may require complex equipment. Methods include chemical indicators in which color changes occur when exposed to specific gases. Physical methods include the measurement of the amount of a specific wavelength of light (usually infrared) absorbed in passing through an air-gas mixture.

The Importance of Air Filtration

Air filtration supplies the means to obtain the level of particulate and molecular cleanliness required by any definition of "air conditioning." It ranges from the simple task of preventing larger particles from plugging heating/cooling coils, to removing particles or molecular contaminants as small as 0.1 µm and smaller which could cause a short circuit on a microchip.

Facility managers should always look to air filtration as the best way to protect the health and safety of the occupants in a facility by removing contaminants from the air. Other reasons include:

- Protect the decor of occupied spaces by removing the staining portion of airborne dust;
- Reduce maintenance of building interiors by reducing the frequency of washing such items as Venetian blinds and fluorescent fixtures;
- Protect the contents of occupied spaces including paintings, tapestries, and other items of historic or cultural value;
- Eliminate fire hazards by removing lint and other materials that might accumulate in ductwork;
- Extend shelf life of perishable products by removing airborne mold and bacteria during processing operations; and
- Remove airborne bacteria from operating room air to help prevent postoperative infection.

Since it can be shown that the air we breathe is frequently contaminated, the logical way to eliminate or reduce this contamination is by the proper application of air filters. Before considering how filtration can solve the problem, we need to define several terms:

Space is a defined area, usually enclosed by walls and a ceiling or the equivalent. The walls may have openings such as doors and windows. The space can be a building, a room, or part of a building or room.

Ambient air (room air) is the air which surrounds the occupant or a process in a space.

Recirculated air is air which has been taken from the space, reconditioned (temperature, humidity, and cleanliness adjusted as necessary), and returned to the space.

Return air system is a combination of ductwork and fan (sometimes omitted) which takes air from the space and delivers it back to the conditioning equipment.

Conditioning equipment is the combination of air filters and other air cleaning devices, heating/cooling coils, and humidifiers that treat the air to be supplied to the space. Air cleaning devices utilizing ultraviolet light are available to control microorganisms. They are discussed further in Chapter 14: *"Ultraviolet Germicidal Irradiation, Photocatalytic Oxidation."*

Outdoor Air (OA) is air taken from outdoor atmospheric air.

Supply air is a mixture of recirculated air and outdoor air that has been conditioned and delivered to the space.

Supply air fan system is a combination of fan, ductwork, and diffusers which delivers supply air to the space.

Diffusers are terminal devices on the supply air fan system. They distribute the conditioned air through the space. Aspirating diffusers mix supply air with ambient air before distributing the mixture through the space. Non-aspirating diffusers discharge supply air directly into the space without mixing with ambient air.

Exhaust air is air taken from the space and not reused therein. Exhaust air is usually ducted to the outside atmosphere.

Exhaust air fan system is a combination of fan and ductwork which removes air from the space and discharges it to the outside atmosphere.

Make-up air is outdoor air supplied to replace exhaust air. An example would be air supplied to a facility in a specific quantity to replace air removed by bathroom or kitchen hood exhaust.

Infiltration is air introduced into the space through openings between the space and its surroundings when the amount of make-up air is less than the exhaust air. The space is then said to be under negative pressure.

Exfiltration is air moved out of the space through openings between the space and its surroundings when the amount of make-up air is greater than the exhaust air. The space is then said to be under positive pressure.

An air conditioning layout including all these components and systems is shown in Figure 1.8. Of all the elements shown, the most important ones from the viewpoint of supply air cleanliness are outdoor air, return air, and filters.

Outdoor air is, in the ideal world, free of all dust and gases that could affect a space's occupants, decor, or contents. In the real world it is rarely if ever deserving of the description "fresh" and must be filtered to remove the contaminants it would bring inside.

Return air contains the particulate material that was generated within the space from processes, people and infiltration.

Supply air, as delivered to the space, is intended to be of the cleanliness necessary to achieve the objectives of the space. However, once this supply air reaches the diffuser, its utilization depends on the design of the air diffusion system. If this clean air is not distributed properly throughout the space, comfort may not be achieved. If it is not used to protect a product or process in the space, its cleanliness may have no meaning. More information is given in Chapter 9: *"Controlled Environments."*

System Filters

The number of filter banks used in a system will depend on the arrangement of the conditioning equipment and the location of the supply fan as shown in Figure 1.8. In normal draw through comfort air conditioning systems, the supply fan is the last item downstream and only Filter #1 is used. This filter is used to protect the mechanical equipment and to provide the degree of cleanliness required for normal occupancy and space usage. Filter #1 may in fact be two separate filters in series. The first, a prefilter, is intended to extend the life of the down-stream second or final filter. In such a system, special concern is necessary to avoid inward leakage of unfiltered air which may occur after the filter and before the fan inlet.

In critical HVAC systems, the possibility of contamination due to inward leakage is avoided by placing the supply fan in the alternative location. All of the conditioning equipment is then under positive pressure and any leakage will be outward. This is normally referred to as a

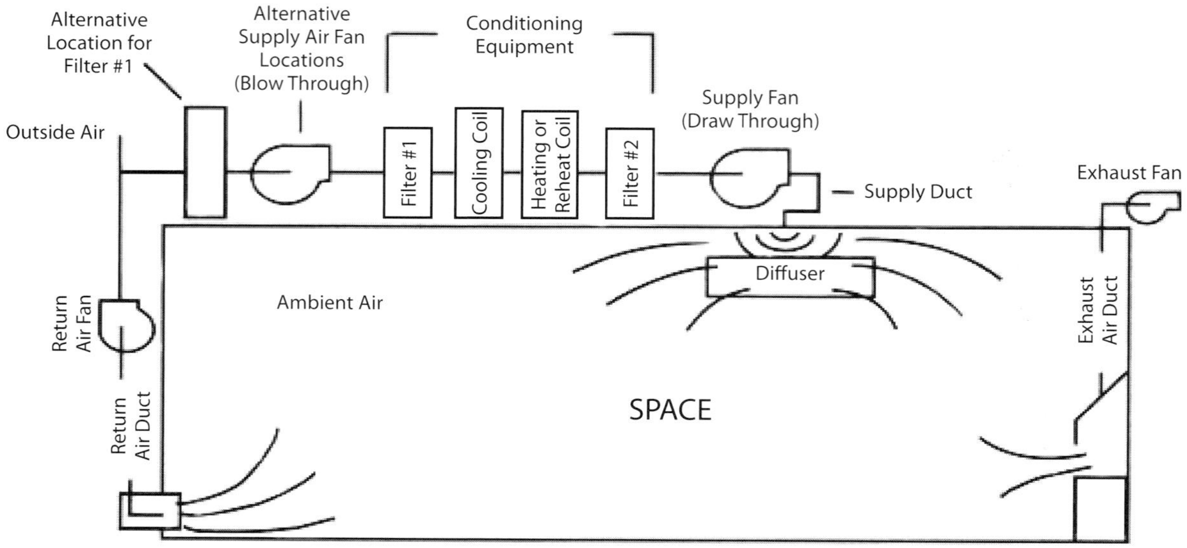

Figure 1.8. Components of an Air Conditioning System

"blow-through system." Filter #1 will protect the conditioning equipment and act as a prefilter for Filter #2. All the air entering the supply duct will flow through Filter #2. Additional filters in a 3-bank system may be located downstream at the space. These filters are called "terminal" filters and can most often be found in cleanrooms and hospital operating rooms.

One practical filtration problem associated with blow-through system relates to the supply fan location. If it is located too close to the filter bank, the fan discharge velocity may be so high that Filter #1 is damaged. Diffusion plates and other devices can be used to dampen the discharge velocity and distribute air uniformly across the filter face. Sometimes Filter #1 is moved to its alternative location.

These simple arrangements of conditioning equipment should not be considered to be the extent of available alternatives. Although they are the basic arrangements, other modifications can be used for special conditions. These are discussed in later chapters.

Filter Efficiency

In the past, filter performance was not given in terms of ability to remove particles of any given size, except for HEPA filters. See Chapter 8: *"HEPA and ULPA Filter Testing."*

But now, air filter performance may be categorized by testing methods developed by ASHRAE, ISO and IEST. The United States government National Bureau of Standards (NBS) particulate test was used prior to 1968. Since then, ASHRAE is the organization that has researched and developed filter performance standards (Standard 52) for heating, ventilating and air conditioning filters.

ANSI/ASHRAE Standard 52.1-1992, which is in the process of being assimilated into ANSI/ASHRAE Standard 52.2-1999, defines filter cleaning ability either in terms of (1) arrestance using ASHRAE synthetic test dust or (2) atmospheric dust spot efficiency.

ANSI/ASHRAE Standard 52.2 defines a filter's fractional efficiency and reports it as a minimum efficiency reporting value (MERV) using a challenge of potassium chloride particles. The particle sizes are sampled upstream and downstream of the filter under test, and counted and sorted into 12 different size ranges. This information is processed and the single-number filter efficiency reported according to final removal efficiency numbers averaged in three size ranges. More detail on filter testing is given in Chapter 7: *"HVAC Filter Testing"* and Addendum 7.1.

It is important to look at all the results from any test report. The first page of the test report will usually give summary information, but it is necessary to look at all of the test data to know how the filter performed from the beginning to the end of the test.

Modeling of Filtration Systems

Efforts to achieve a certain level of cleanliness within a space depends on much more than filter efficiency alone. Air distribution and air quantity are equally important. Mathematical and computer modeling can show the effects of varying supply air cleanliness, air quantity, and air distribution. This can become complex. However, it points out ways to improve system design so that all three factors are optimized. Since this book is concerned with air filtration and air filtration systems, the major point to be made is that achievement of a level of cleanliness within a space does not depend on filtration alone. The quantity of air and its distribution within the conditioned space are also very important.

Notes to Chapter 1

1. Air quality criteria for particulate matter, EPA, 1996.
2. 2005 ASHRAE Fundamentals Handbook, American Society of Heating, Refrigerating and Air Conditioning Engineers 1719, etc., page 12.3.
3. International Organization for Standards, 2002, Cleanrooms and Associated Controlled Environments 14644-1.
4. *An Evaluation of Atmospheric Dust*. American Air Filter Bulletin AF-1-108.

2

The Principles of Air Flow, Air Pressure and Air Filtration

Principles of Air Flow

The flow of air between two points is due to the occurrence of a pressure difference between the two points. This pressure difference results in a force placed on the air, usually by a fan, causing the air to flow from the area of higher pressure to the area of lower pressure. The quantity of air, usually referred to in cubic feet per minute (CFM) is represented by the symbol Q. The speed of flow or velocity of the air usually referred to in feet per minute (FPM) is represented by the symbol V. The size of the conduit through which the air flows, usually ductwork, is referred to as area expressed in square feet and is represented by the symbol A.

The air flow through a conduit – ductwork or a filter is expressed by the formula:

$$Q = VA$$

Therefore, to find one of the other components, the formulas would be:

$$V = Q/A$$

$$A = Q/V$$

An example of the application of this formula would be a 24″ x 24″ x 2″ flat panel filter (A = 4 square feet) placed in an airstream of 2000 cfm (Q=2000) would yield a velocity through the filter (V) of 500 fpm (2000/4=500).

Principles of Air Pressure

As air travels through a conduit, it creates a pressure called velocity pressure (VP). There is a relationship between velocity of the air and the velocity pressure based upon the density of the air. This relationship can be expressed in the formula:

$$V = 4005\sqrt{VP}$$

Where:

V = velocity in feet per minute

4005 = standard density of air derived from gravitational acceleration (32.2 ft/sec^2 and air density of 0.075 pounds per cubic foot)

VP = velocity pressure in inches of water

Velocity pressure is measured in the direction of flow through a conduit and is always positive.

Air confined in a conduit whether in motion or not creates another type of pressure which exerts itself in all directions at the same time. Sometimes referred to as "bursting pressure," this pressure is called static pressure (SP). Static pressure is independent of the velocity of the air and can either be positive or negative depending on where it is measured in the conduit. This pressure can be measured using

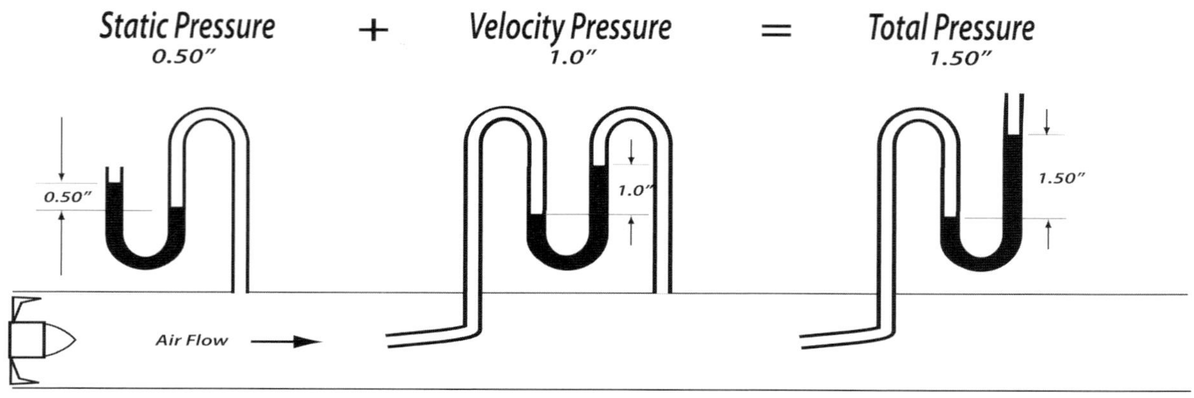

Figure 2.1

a device known as a Pitot (pronounced pea-toe) tube. Figure 2.1 shows the relationship of velocity and static pressure and the relationship of these pressures is expressed in the formula:

$$TP = SP + VP$$

Where:

TP = Total pressure

SP = Static pressure

VP = Velocity pressure

To understand the pressure relationship, the lowest pressure in a system is at the fan inlet (A) and the highest pressure is at the fan outlet (B) see Figure 2.2. For example, as a filter loads, the resistance to flow causes an increase in the pressure on the upstream side of the filter and a resultant decrease on the downstream side.

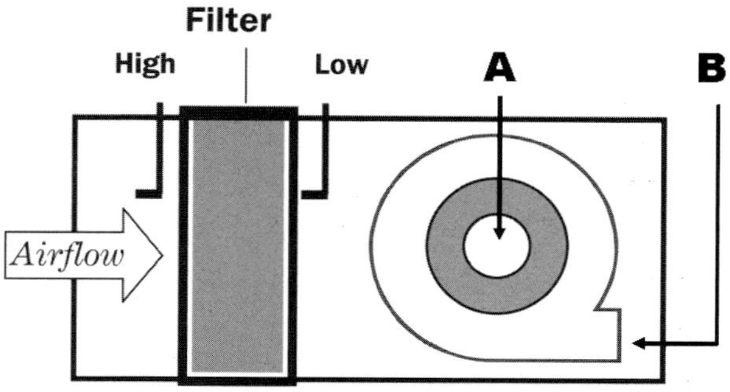

Figure 2.2. Typical Location of Pressure Drop Measurement Devices

This drop in pressure between upstream and downstream is referred to as "pressure drop."

Filter pressure drop measuring devices usually measure static pressure increases in the system created by a filter as it loads. This pressure is expressed in inches of water gauge (Figure 2.3), i.e., the pressure it takes to raise water 1″ in a tube..

As a filter loads in a system, depending on the type of fan and its place on the fan curve, velocity of the air decreases as static pressure increases causing the system to work harder to deliver air to the space. This increase in work by the fan results in increased energy consumption and higher owning and operating costs (see Chapter 13: *"Owning and Operating Costs"*). Figure 2.4 shows the effect of this pressure in the system.

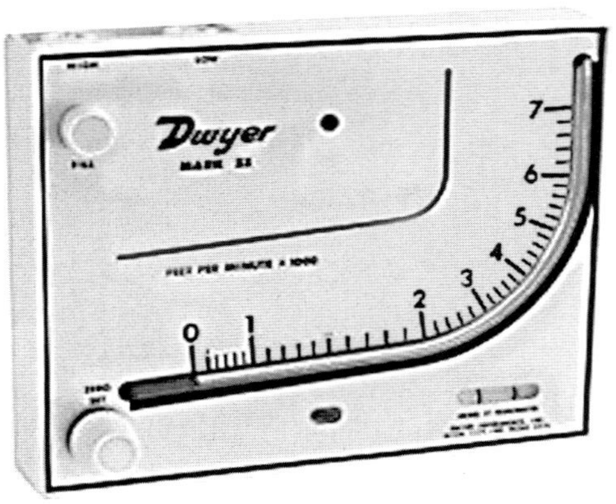

Photo Courtesy Dwyer Instruments

Figure 2.3. Pressure in a system measured by an inclined tube manometer filled with red oil

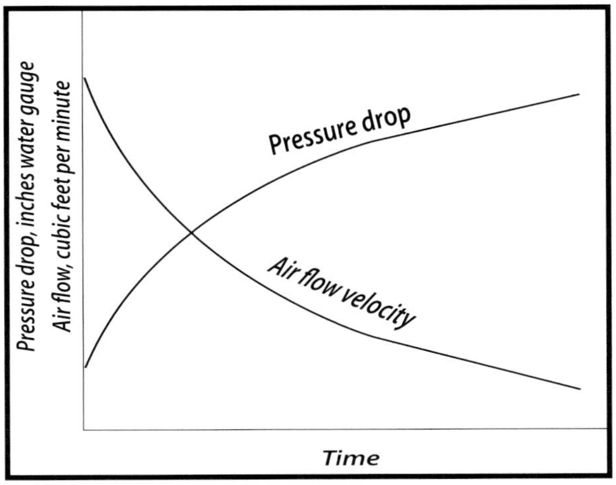

Figure 2.4. Pressure Drop Graph

Principles of Particulate Air Filtration

Air filters are devices that remove aerosols from an air stream as the particulate contaminated air passes through them.

Three Types of Air Filter Categories

There are three types of air filtration categories: mechanical air filters, filters incorporating electrostatically charged filter media and electronic air cleaners.

I. *Mechanical air filters* capture particulates on the filter media[1], the material that comprises the filter element. This capture involves two different considerations. The first is the probability that a particle will collide with or be removed by the "fibers" which make up the filter media. (The word "fibers" is used in the broadest sense to cover any component of a filter media.) The second is the probability that the particle, once contacting the filter will continue to adhere to the fiber.

II. *Electrostatically Charged Filter Media (passive and active)* has been used for several decades. The advantage of charged filter media is that the charge on the fibers increases initial filtration efficiency without affecting resistance to airflow. Particles have a natural charge, or pick up an electrical charge as they pass through the air. These particles, in turn tend to stick to filter fibers. These materials may be used on stand-alone filters or may be combined with other technology to enhance their performance.

III. *Electronic Air Cleaners (two-stage)* are externally powered devices that impose a charge on airborne particles in the "ionizer section". The charged particles are then collected on oppositely charged plates in the "collector section".

I. Mechanical Air Filters

There are four different processes responsible for the capture of particulates in a mechanical filter. Usually one prevails in a specific filter, but rarely is it the exclusive mechanism. These functions are:

- Impingement
- Interception
- Diffusion
- Straining

Impingement is the mechanism by which large, high-density particles are captured. As air flows through a filter, it must bend or change direction many times to flow around the filter fibers. Because of their inertia, larger particles resist change in direction and attempt to continue on in their original directions. For this reason, they collide with, and adhere to the fibers (Figure 2.5).

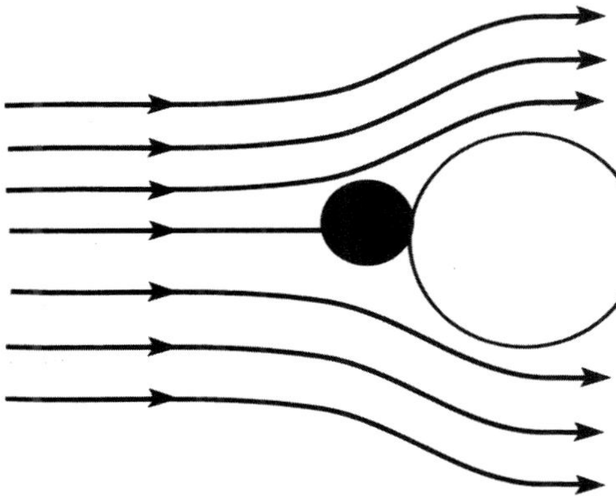

Figure 2.5. Impingement

Interception occurs when a particle follows the air stream, but still comes in contact with the fiber as it passes around it. If the forces of attraction between the fiber and the particle are greater than the force of the airflow to dislodge it, the particle will stick to the fiber. Interception is enhanced when the size of the fiber is closest to the size of the particle (Figure 2.6).

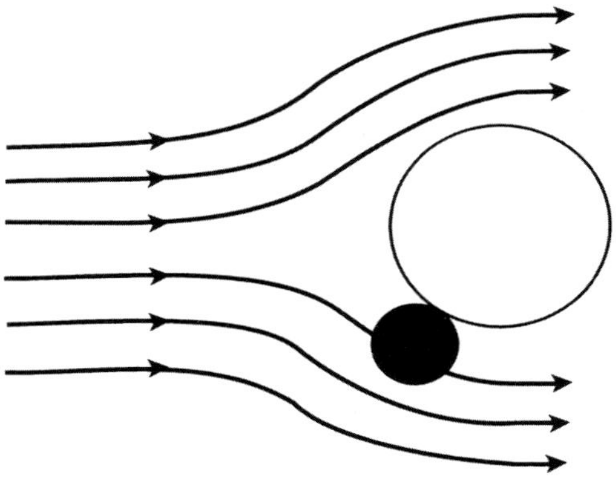

Figure 2.6. Interception

Diffusion explains the capture of very small particles at lower air velocities. As the contaminated air passes through the filter media, minute particles will tend to move from areas of higher concentration and will take an erratic path described as Brownian Motion. This erratic path increases the probability that particles will come in contact with fibers and will stay attached to them. Diffusion works best with fine filter fibers and very low air velocities (Figure 2.7).

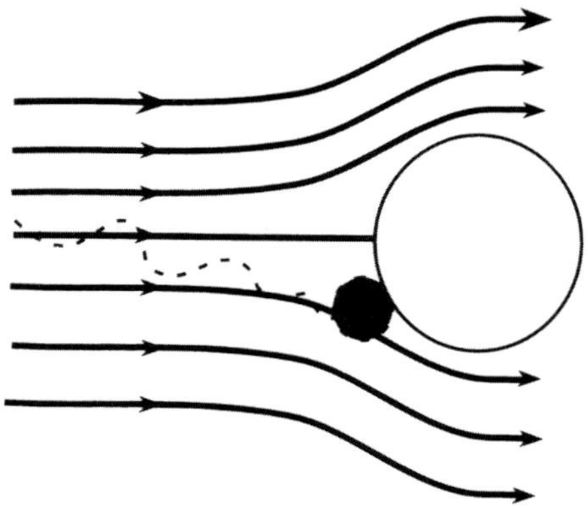

Figure 2.7. Diffusion Effect

Straining occurs when the smallest dimension of a particle is greater than the distance between adjoining filter media fibers (Figure 2.8).

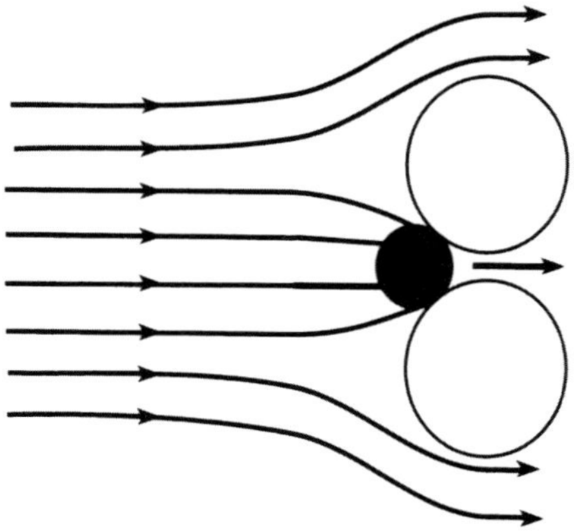

Figure 2.8. Straining

Impingement Filters

The effectiveness of the impingement process depends on the following:

Particle size: The larger the particle, the greater the mass and the greater the possibility that it will have enough inertia to resist change in direction and collide with a fiber.

Particle density: The greater the density of a particle, the greater the mass and the greater its inertia.

Depth of the media and orientation of fibers: The thicker the filter media and the closer the orientation of fibers, the greater the possibility of collision by a fiber.

Velocity of airflow through the filter. The greater the velocity of air through a filter, the greater the kinetic energy of the particles in the airstream, and the greater the inertia of the particles. Overrating and underrating filters not only impacts the pressure drop and service life of a filter but it can also impact filter efficiency. If a panel filter is underrated, its overall efficiency on larger particles may actually be reduced. This is because most lower MERV filters utilize impingement as the main capture

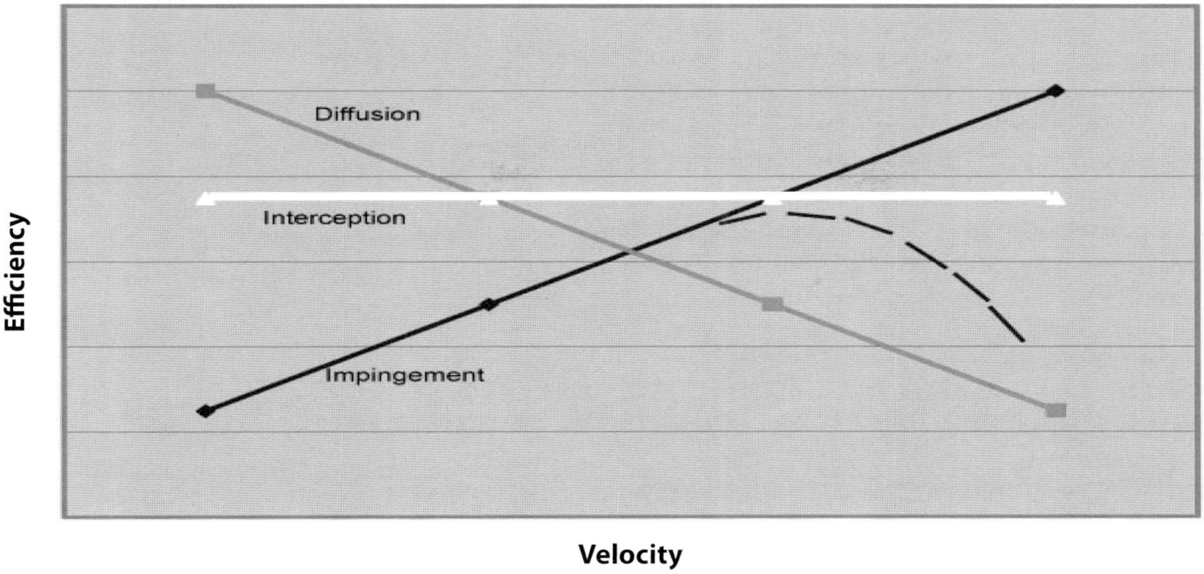

Figure 2.9. Theoretical Effects of Impingement, Interception and Diffusion on Filter Efficiency

mechanism. Reducing the airflow decreases the velocity. Decreasing the velocity will decrease the impingement process by reducing the kinetic energy of the particle in motion. This may increase the capture of smaller particles by diffusion and interception. Conversely, overrating an impingement filter by increasing the airflow may increase its overall efficiency on larger particles (see Note) by increasing the impingement process but decrease the capture of smaller particles. For additional information about overrating and underrating, see page 13.5.

NOTE: Lower MERV filters that predominately utilize the impingement capture principle, experience efficiency decreases at a given point as velocity increases and turns downward (see dotted line Figure 2.9) resulting in lower efficiencies even on larger particles. Filters MERV 12 and higher do not usually experience this phenomenon because they utilize the principles of interception and diffusion as the predominant capture principle. Consult manufacturer recommended flow rates and values before underrating or overrating any filter.

Use of Adhesives

The forces between a filter fiber and a dust particle captured by impingement are relatively weak (see van der Waals forces). There is a good possibility that the particle may become dislodged from the fiber by the velocity of the air passing around the fiber or by system vibration. In order to overcome this possibility, impingement filters are frequently treated with adhesives that coat the fibers and create a bond between them and any particles that may impinge upon the fiber.

The ideal adhesive:

1. Fireproof or fire-resistant
2. Maintains its tackiness between filter changeouts
3. Nontoxic
4. Odorless
5. Non-migrating, meaning it will not entrain into the airstream.

Interception and Diffusion Filters

Interception poses a stronger influence with larger particles, whereas diffusion explains the capture of smaller particles. Both are intended to cause a particle and a filter fiber to come in contact. The strength of the bond between the fiber and the particle depends on several forces of attraction.

van der Waals forces: The perfect gas law states that if the pressure on a gas is doubled, its volume should be cut in half. Researchers found that when gases were subjected to high pressures, the volume reduction was greater than expected. Some force drew the gas molecules together. A theory was developed by Johannes van der Waal based on molecular attraction to explain why this occurred. This same force of molecular attraction helps keep a dust particle attached to a fiber. The original equations developed as a result of studies by van der Waal have been modified by a series of researchers. However, these forces still carry his name.

Dipole moment: Most molecules, even when they carry no charge, have a dipole moment. This happens when there is a greater concentration of electrons at one point or end of a molecule than there is at another. This is especially true of large molecules. As a result, when a particle (even though its charge is electrostatically neutral) comes in contact with a fiber, it tends to align itself so that the more negatively charged side of the particle is adjacent to the more positively charged sectors of the fiber. Neither the particle nor the fiber has to be charged for this to occur (Figure 2.10).

Figure 2.10. Dipole Moment

Disruptive Forces

As was mentioned for the impingement filter, there are forces that tend to disrupt the attraction between a particle and a fiber. The two most important forces are:

1. The flow of air through the filter which, because of the particles drag force, may re-entrain it; and
2. System vibration which may dislodge the particle. Normally the effects of vibration can be minimized by proper filter installation, including vibration isolation.

Filter Equilibrium

In most filter systems equilibrium exists between the forces of adhesion holding a particle to a fiber and the drag force of air on the particle. There will be minimum re-entrainment when the forces of adhesion are equal to or greater than the drag force. A mathematical model for this equilibrium is given in Appendix Four.

Factors Affecting Interception and Diffusion

Particle capture by interception and diffusion is affected by the following factors:

Size of particle: Larger particles are more likely to be captured by impingement. Smaller particles, on the other hand, have an enhanced opportunity of coming in contact with a fiber because of their wider effective path created by Brownian Motion and media velocity. [There is a most penetrating particle size (MPPS) for any high efficiency filter operating at a specific media velocity.]

Fiber diameter: The strongest force occurs between particles and fibers that are approximately the same diameter. Consequently, the smaller the filter fiber, the greater the retention of smaller particles.

Depth of the media and orientation of fibers: The thicker the filter media and the closer the orientation of fibers, the greater the possibility of capture by a fiber.

Mixed-Type Mechanical Filters

Much work has been and continues to be devoted to determining the contribution of different mechanisms of filtration to the overall performance of different media and to developing a more precise mathematical formula to describe their performance. Rarely is an exclusive mechanism responsible for all the collection observed. Normally however, impingement is the predominant influence in filters intended to capture large particles traveling at higher velocities, and interception and diffusional effects are the main influences in the capture of smaller particles traveling at lower velocities.

Factors Affecting Mechanical Filter Selection

The following considerations are involved in the selection of a mechanical air filter:

Efficiency: The most important consideration is the ability of a filter to remove from an air stream the greatest number of particles. This ability is described as efficiency and will be discussed in Chapter 7: *"HVAC Filter Testing."*

Pressure drop: The resistance to airflow created by an air filter is an important consideration. The higher the resistance, the greater the energy required to overcome it. Consequently, all other considerations being equal, the filter with the lowest pressure drop is preferred.

Capacity: This is the amount of air that a filter can handle. Usually capacity is defined as the volume of air per unit of time that a clean filter can handle at a specified pressure drop. This volume is expressed in cfm (cubic feet of air per minute) or in SI as m^3/s (cubic meters of air per second). Changing the amount of air being handled by a filter will affect other performance values such as pressure drop and efficiency.

II. Electrostatically Charged Media

Electrostatic attraction (passive): In addition to active electrostatically charged material discussed in the next section, it is also possible for filter media to become electrostatically charged by the flow of air (especially dry air) through it. Filters incorporating such media are described as "passive electrostatic filters." Most particles are charged naturally, and are held by strong electrostatic forces to the oppositely charged fiber with which they come in contact (Figure 2.11). The smaller a particle or a fiber, the relatively stronger the electrostatic forces will be.

Electrostatic attraction (active): Synthetic filter fibers can be actively charged during manufacture to be either positively and/or negatively charged and then made into a non-woven filter material called "electret" media. Charged fiber technology can be classified by the method used to create the electrostatic charge:

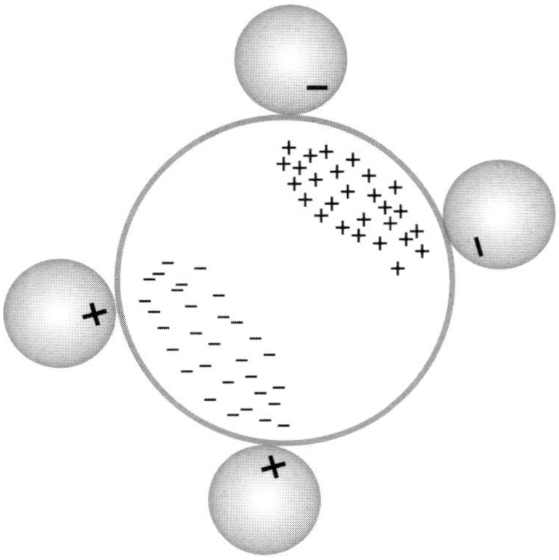

Figure 2.11. Particle Attraction to Charged Fiber

1. Triboelectric charging
2. Corona charging
3. Charging by induction

Triboelectrically charged material results from the rubbing together of dissimilar polymers. The first use of this technology involved wool with a resin treatment. Wool is an excellent conductor and the most electropositive fiber. Resin is an extremely good insulator and when combined with wool and rubbed together will exchange charge with the wool and become negatively charged. Thus the resulting product contains positive and negative charges that will attract and hold particles of the opposite charge. More recently triboelectric-charged material has been developed by blending together two fibers of dissimilar characteristics. The rubbing together of these fibers will cause an exchange of electrons resulting in one material developing a positive charge and the other developing a negative charge. The most common materials currently used for this product are polypropylene and modacrylic. Blends of other binary dissimilar fibers will also result in triboelectric properties.

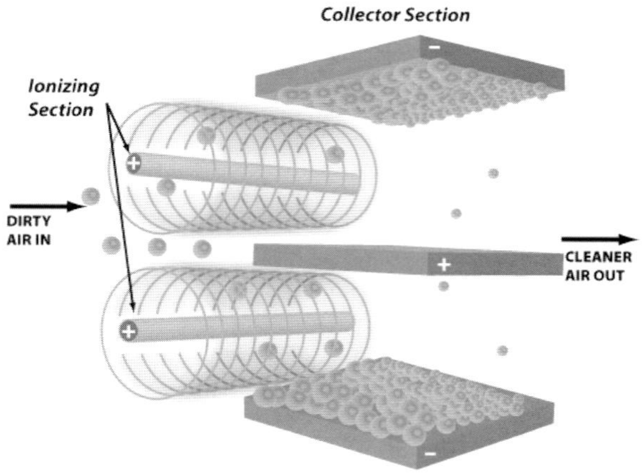

Figure 2.12. Animation of EAC Collection System

Corona charged material involves exposing fibers or filter material to an electrode designed to create high voltage either positive or negative. The ions generated by the electrode are collected on the surface creating the static charge. Split-fiber technology involves extruding a sheet of polymer (polypropylene), stretching it to create a molecular alignment and permitting the sheet to be split into fibers by a process called fibrillation. The entire sheet is subjected to positive charges on one side and negative charges on the other. The resulting fibers therefore have both positive and negative charge.

Charging the finished material as a whole is a second method of corona charging. The corona charge is applied to the surface of the finished material resulting in a much simpler process but also results in a product with a lower charge density.

Material charged by induction involves a process similar to that used in the production of electrically charged sprays. Fibers produced by electrostatic extrusion are typically quite fine and indicate presence of both charges. Filters produced from fibers that have been charged by induction are generally quite efficient due to their mechanical capture efficiency complimented by their electrostatic action. However, this method has not proven as effective as either triboelectric or corona charged material.

While this additional charge increases the forces that attract and hold particles, they may not last the life of the filter. Depending on the type of charging technology, other factors such as humidity, time in storage and dust loading may erode or blind the electrostatic charge.

III. Electronic Air Cleaners Reference Chapter 6

Electrostatic Precipitation

In the first stage (ionizing section), the particles in an air stream are given an electrostatic charge. In the second stage (collecting sections), these particles are removed from the air stream by electrostatic attraction to oppositely charged plates (Figure 2.12). In earlier designs, a 12 kV charge was placed on the ionizer section to electrostatically charge the particles. This section was made small and was designed with ionizing wire and grounded struts so that ozone generation was minimized. In the collecting section, the alternately charged plates were sized and spaced so that a 6 kV charge was enough to collect the particles but not enough to create ozone.

Photo Courtesy Honeywell

Figure 2.13. In-duct Residential Electronic Air Cleaner

Devices used in air-conditioning and ventilating systems are now called electronic air cleaners to distinguish them from the earlier high-voltage, stack gas cleaning electrostatic precipitators. Electronic air cleaners are discussed in detail in Chapter 6: "*Air Cleaners.*"

Notes to Chapter 2

1. The material in a filter that captures particles is the filter medium. The plural of medium is media. However, there is a common practice in the air filter industry to use media for the singular and medias for the plural form. Both conventions are used in this book but hopefully create no confusion as to whether singular or plural is intended.

3

Impingement Filters

Definition

Panel filters are mechanical filters whose principal mechanism of particle capture is impingement. See Chapter 2: *"Principles of Air Flow, Air Pressure and Air Filtration".*) They are characterized by relatively high media velocities (the velocity of the air passing through the filter media) and by low pressure drop. They are primarily intended for the capture of heavy concentrations of relatively coarse dust particles. Many filters in this category also depend on an adhesive coating, called a tackifier, on the media to retain the captured dust.

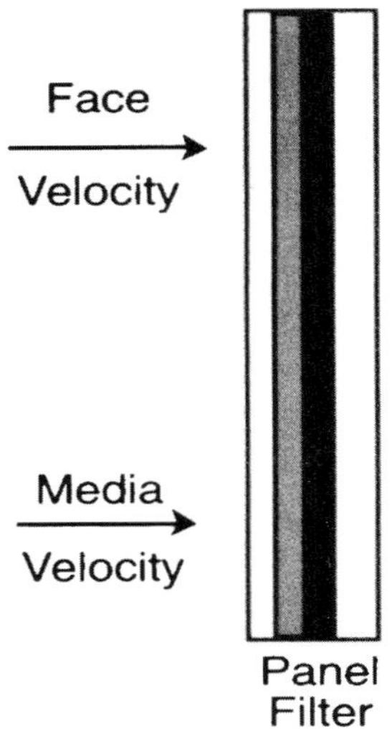

Figure 3.1 Face Velocity is equal to Media Velocity in a Panel Filter.

General Categories

Impingement filters can be categorized in several different ways, but for convenience we will consider them by their configuration:

- Panel filters
- Washable metal-panel & baffle grease filters.
- Roll filters
- Lint filters—special applications of roll filters

Panel Filters

Panel filters are media air filters in which the face velocity and media velocity are the same (Figure 3.1). The fact that both velocities are the same requires that the media be in a flat panel form.

Panel filters can be categorized by the material used as the filter media, and the method of servicing.

Materials of Construction

Metals are used as filter media where heavy-duty usage is involved and where filters are intended to be washed. Steel and aluminum are the most frequently used materials, however, stainless steel or monofilament polypropylene filters are often used where corrosion is a concern.

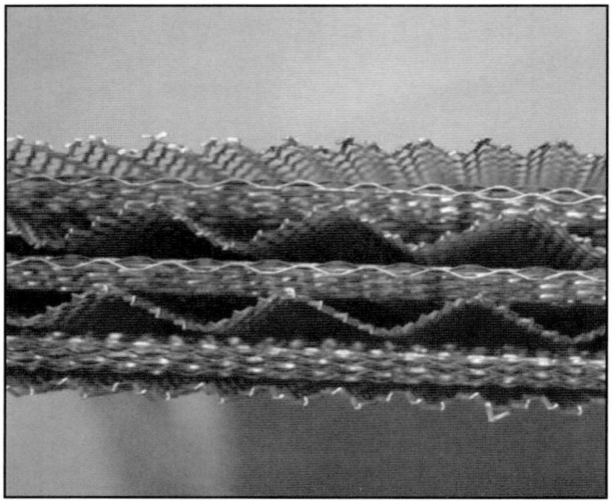

Photo Courtesy Smith Filter

Figure 3.2. Filter with Graduated Density Media

Many metal filters use graduated density media, i.e. the upstream side is a more open material while the downstream side is more densely packed. The purpose is to allow a full dust loading of the filter by the capture of large particles on the upstream face and progressive capture of smaller particles as the air moves through the filter (Figure 3.2).

A second category of metal filters is the formed-screen type. They are made by crimping strips of media (usually screens) to create a highly porous filter element. As dirty air flows through the media, the dust impinges on the screen wire. Units of this design (Figure 3.3) have high capacity and low pressure drops.

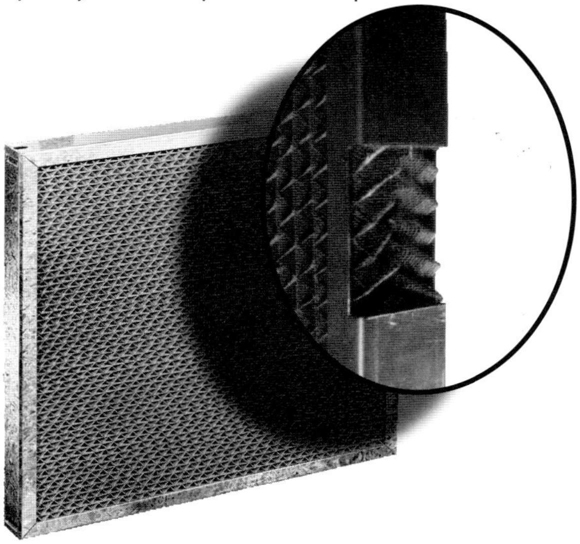

Photo Courtesy Airsan

Figure 3.3. Crimped Screen High Velocity Filter

Spun-glass fiber media is used extensively in impingement filters. The fibers are usually adhesive-coated to improve retention. Fiber diameters are in the 18 to 30-μm range. The density of the media can be varied during manufacture. The thickness of spun-glass pads can vary from 3/8 in. (9.5 mm) for room air conditioning units to 5 in. (127 mm) for certain heavy-duty transportation applications. The 2 in. (51 mm) thick filter is used for most commercial-industrial applications. Spun-glass pads can be used by themselves in metal retainers or they may be incorporated into cardboard filter frames (Figure 3.4). Pads or filters incorporating spun-glass pad material are discarded after use.

Photo Courtesy Purolator

Figure 3.4. Throwaway Spun-glass Media Filter

Synthetic fibers are also used as the media in panel filters. Several kinds of synthetic fibers are used for this purpose. Polyester is the most common material, but other fibers such as polypropylene, nylon, and modacrylic are used in special applications.

Fiber size is measured in denier (pronounced den-yer). Although denier is, in fact, a measure of weight, it can be used to indicate fiber diameter. For example, a 6 denier polyester fiber is 25 μm in diameter. A 40 denier polyester fiber is 66 μm in diameter.

One type of media commonly used is called "dual denier." It incorporates a larger (25 to 40 denier) fiber on the air entry side and a smaller (2- to 15-denier) fiber on the air exit side. The intent is that the air entry layer captures the larger particles and the air exit layer captures particles too small to be caught by the large denier fibers.

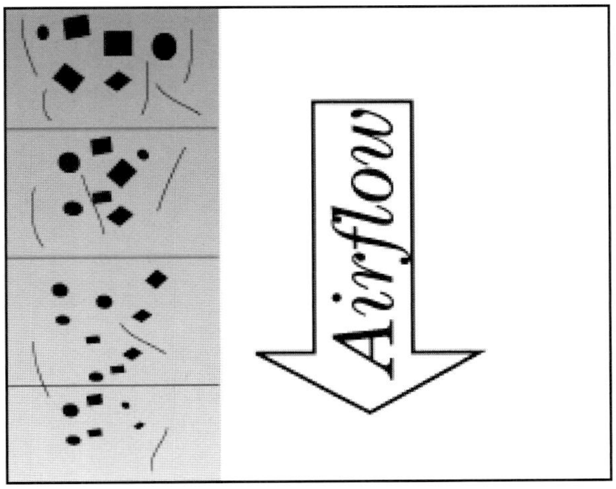

Courtesy Tri-Dim Filter Corporation

Figure 3.5. Visual of depth-loading of filters

In making air filter media with synthetic fiber the most important consideration is consistent fiber distribution. Regardless of the method used, the end product must be free of voids, thin spots or lumps of fiber. These conditions can promote uneven loading of particles on the media or increase the chance of particles migrating through the media and exiting of the downstream side of the filter.

Tackifiers are sometimes used with synthetic air filter media for flat panel and automatic roll type filters. They serve the same purpose as adhesives on glass and washable media. A tackifier can be an adhesive material which softens the mat and makes it sticky or it can be an oil. Tackifiers are typically applied to the air exit side of synthetic media to enhance dust holding and prevent particle migration.

Air filter media incorporating synthetic fibers may also have a color applied to indicate the entry or exit side. The color used in this type of media has no bearing on its efficiency.

Air filters using synthetic fibers are usually designed to be discarded at the end of the filter's useful life.

Larger monofilament fibers made from polypropylene and similar materials called "passive electrostatic" filters are woven into media. Filters incorporating these plastic fibers are usually intended to be washed when they have become loaded with dust. They are usually used without adhesives. Part of the dust-collecting characteristics of some plastic filters is attributed to the electrostatic charge that may be generated as air flows through them. For example, polyester has a positive charge and polypropylene a negative charge. Airborne particles are attracted to fibers that have a charge opposite their own. This attraction is diminished when the air is humid and as particles build up on the fibers.

They are called "passive" because they are "self-charging" as compared to electronic air filters that use an external power source, or electret media which is charged during manufacture.

Foam material, usually open-cell polyurethane, is usually 1/4 to 1/2 in. (6.4 mm to 12.8 mm) thick and is used for very low velocity applications. Foam materials are classified by pore size, ranging from 10 to 100 pores per in. (2.5 to 25 pores per cm). The greater the number of pores per inch, the higher the efficiency and pressure drop of the media. Foam materials can be washed and reused, but tend to break down over time.

Media Performance

Typical performance values of different impingement filters and media are shown in Figure 3.6. Since these are only typical values, information about the performance of a specific filter should be secured from the filter manufacturer, who frequently can provide an ANSI/ASHRAE Standard 52.2 test report giving the filter performance. Test results on filters purchased on the open market by an independent testing laboratory will normally show typical performance values.

Figure 3.6. Typical Performance Values of Impingement Filters

Media Description	Face Velocity fpm (m/s)	Clean Pressure Drop in. w.g. (Pa)	ANSI/ ASHRAE Standard 52.2 MERV	Final Pressure Drop in. w.g. (Pa)
2" washable metal, graded density, adhesive coated	350 - 500 (1.78)	0.06 (14.95)	4	0.50 (124)
1" washable metal, graded density, adhesive coated	350 (1.78)	0.05 (12.45)	4	0.50 (124)
2" formed screen permanent, washable	300 - 500 (1.52) (2.54)	0.10 (24.91)	4	0.50 (124)
1" thick spun-glass	300 (1.52)	0.06 (14.95)	4	0.50 (124)
2" thick spun-glass	500 (2.54)	0.08 (19.93)	4 - 6	0.50 (124)
1" thick dual-denier polyester	300 (1.52)	0.14 (34.87)	4 - 8	0.50 (124)
0.5" thick monodensity polyester	300 (1.52)	0.10 (24.91)	4	0.50 (124)
1" thick monodensity polyester	350 (1.78)	0.08 (19.93)	4 - 7	0.50 (124)
Passive electrostatic- polypropylene media	300 (1.52)	00.12 (29.89)	4	0.50 (124)

Notes:
(1) These are typical values.
(2) Filter manufacturers should be contacted for performance data of specific filters

Method of Renewal

Filter banks incorporating panel filters are renewed by either washing the filter or replacing the filter and/or filter media. The use of disposable materials has become the predominant method of renewal.

The method of washing filters depends on their construction and the material used. Back-washing (the flow of washing solution in the opposite direction of the airflow) metal or plastic filters is used for most effective cleaning. Washing tanks are sometimes used. The washing solution may, depending on the type of filter media, contain detergents and alkali of different types.

Regardless of the method of washing, the filters should be allowed to drain dry if adhesives are going to be applied to them. Early adhesives for washable filters were refined petroleum-based oils. Newer materials are emulsions that leave a tacky surface on the media when the water from the emulsion evaporates. It is important that the adhesive be applied adequately to coat all the fibers. Filters coated with adhesive should be allowed to drain off excess adhesive before they are put back in service.

Filter Holding Systems

Framed washable and disposable panel filters are usually clamped into metal holding frames. These frames are usually either painted or galvanized steel of sufficient metal thickness to provide the support needed by the filter. The

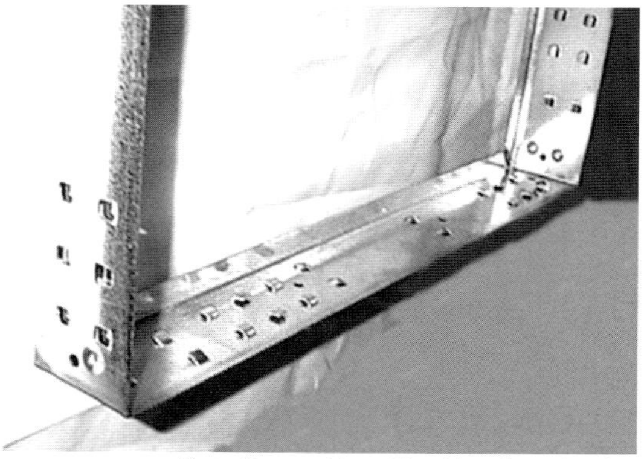

Photo Coutesy Aircon Filter Manufacturing, Inc.

Figure 3.7. Filter Holding Frame

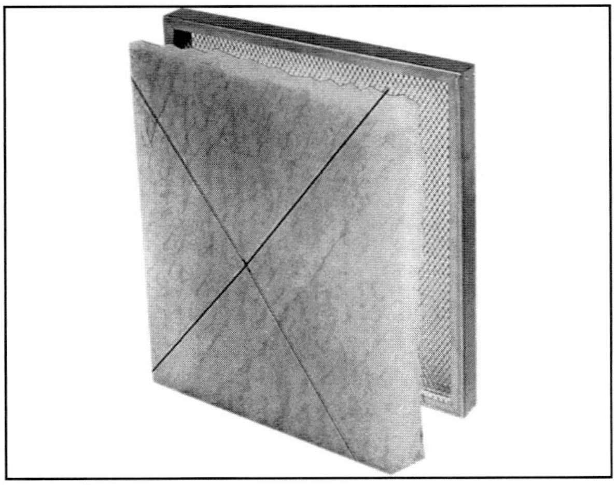

Photo Courtesy Tidewater Air Filter Fabrication Co., Inc.

Figure 3.8. Filter Pad with Holding Frame

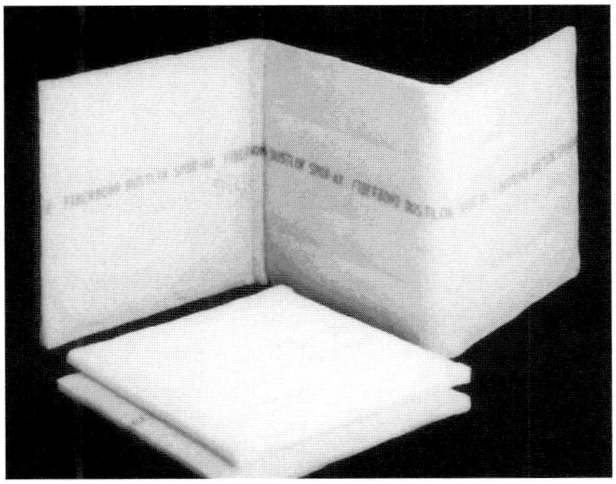

Photo Courtesy Fiberbond Corporation

Figure 3.9. Ring and Link Panel Type Air Filters

frames are assembled into banks of suitable size to meet the needs of a system (Figure 3.7). The space between holding frames and around the perimeter of the filter bank should be caulked to prevent the by-pass of any unfiltered air. Large banks should be reinforced with suitable vertical stiffeners. When pads of spun glass or synthetic material are used, a backup grid is added to the downstream side of the holding frame to support the mat when the system is in operation. A gate or equivalent device is used to support the upstream face of the pad between changes (Figure 3.8).

Ring Panels and Link Panels

A design incorporating a wire ring sewn, thermally sealed or chemically bonded inside a panel made from two or more layers of media (Figure 3.9) is called a ring panel. Ring panels are manufactured from rolls of media, and then cut into separate panels. If the rolls are not cut into separate panels, they are referred to as a link panel and have multiple panels in each link. The ring panel and link panels are sized so that the filter creates a friction fit when installed into its holding frame. The ring eliminates the need for modification to an existing holding frame, and the ring panel seals the edges to prevent air by-pass. This design is also useful when filters of non-traditional sizes are required for certain equipment. An alternative design called a sleeve, consists of layers of media sealed together into which a wire ring can be slipped. The ring may be reused when the media sleeve becomes loaded with dirt and is discarded.

Air handling units incorporating filter sections are frequently supplied without individual holding frames. Framed filters are installed in metal supports where they are positioned side by side. Air by-pass frequently occurs if the filter frames are not tightly fitted against one another or if there are voids between the last filter and the walls of the filter housing section.

Side access housings (Figure 3.10) permit panel filters or link panels to be loaded and removed from one or both sides of the housing. The filters usually slide into a set of top and bottom channels that may be gasketed to prevent top or bottom leakage of unfiltered air. The vertical space between filters is eliminated by making them fit closely against one another and by adding 1/16 inch gasketing to the vertical side of the filter frame.

Photo Courtesy Filtration Group, Inc.

Figure 3.10. Side access housing for Panel Filters

It is important that there be no voids between the last filter installed and the closed access door of the housing. This can be accomplished by means of a spacer – usually a galvanized light-gauge piece of metal, sized to fit in the row and pleated to form a semi-flexible strip that, when inserted and the access door closed, will exert pressure against the vertical side of the filters to prevent air bypass.

Automatic Roll Filters

Ed: Note: This type of air filter is limited in production, but is still in use today in many applications. Information is included here as reference – (see also NAFA text book: Installation, Operation and Maintenance of Air Filter Systems, 2nd Ed.).

Roll filters are devices in which clean filter media is unrolled at one end, exposed to a dirty air stream, and advanced at intervals to keep the pressure drop through the exposed air stream within a desired operating range. Dirty media is rolled onto a take-up reel at the opposite end of the filter. When a roll of media has been completely used and is wound on the dirty media reel, this take-up reel with the dirty media is removed and replaced with the empty reel that held the clean media. A roll of clean media is installed, and the clean media drawn down through the filter space and attached to the empty take-up reel (Figure 3.11).

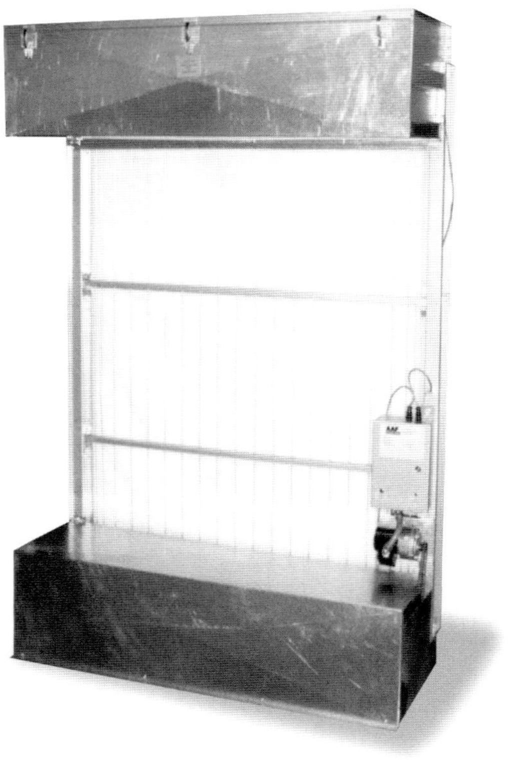

Photo Courtesy American Air Filter

Figure 3.11. Vertical Roll Filter

Roll filters come in either a vertical or horizontal configuration. Vertical roll filters come in nominal 3, 4, 5, and 6 ft. (.91, 1.22, 1.52, and 1.83 m) width modules. Their heights vary with the quantity of air to be filtered and the open space that they must fill. The height variation is usually in increments of 1 ft.. (0.305 m). Clean media rolls are installed at the top of the filter and dirty media at the bottom. Modules can be combined to provide the necessary width for vertical units.

Horizontal filters have heights corresponding to vertical filter widths, and widths corresponding to heights.

Media Advancing

In the automatic roll filter, the unrolling of clean media and the rolling of dirty media is accomplished by motor-driven gear reducer systems. Systems are available so that one drive can move media in several modules. Drives may be located out of the air stream. When a drive must be located in an air stream, its preferred location is on the downstream (clean air) side.

Media movement in a roll filter can also be accomplished by a manually operated chain drive, connected to the dirty media spool. In most instances, there is a manual drive for each module. Manual filters are frequently advanced so that when the media has become loaded, a full new face of it is exposed to the air stream.

Roll Filter Media

Spun-glass media used in an automatic roll filter has a coating of adhesive on the fibers which assists in retaining particles once they have been captured. It also usually incorporates a backing material of some type to give it the necessary strength to be pulled by the drive system without being damaged and to prevent it from bowing or necking. In those instances where required by fire regulations, the media may be used without an adhesive coating. Dry glass media is not as efficient as coated media.

Roll filter media incorporating synthetic fibers are more prevalent and can be used in place of glass. Polyester is the most popular material.

Special Particulate Removal

Lint Filters

Filters are available for lint removal using the automatic roll filter principle. One of the backing materials used for glass media filters is a spunbond nonwoven synthetic. This can be used by itself in a modified automatic roll filter to capture lint and other fibrous materials such as ink mist in a newspaper printing room.

Grease Filters

Grease baffle-style filters are devices installed in commercial kitchen range hoods intended to prevent the accumulation of grease in the exhaust ducts from the hoods. All filters used for this service are made from metal, usually steel or aluminum, and are constructed in such a way as to provide a torturous path through the filter on which grease vapors can impinge and airborne grease splatter can be collected (see Figure 3.12). They are installed at an angle that allows captured grease to drain down the filter and out through holes in the frame. From there, the grease runs down the filter track and into a sump. Grease filters are not effective in removing cooking smoke.

Baffle filters must meet the requirements of NFPA Standard 96 2004. Listed grease filters are tested and certified by a nationally recognized test laboratory under UL Standard 1046.[1]

Older model grease filters are made of expanded metals called metal mesh filters. Because they capture and hold the grease as opposed to allowing it to drain off, they are a hazard when loaded. **Mesh filters no longer meet the requirements of UL Standard 1046 and, therefore cannot be used as primary grease filters.**

Photo Courtesy Smith Filter

Figure 3.12. Baffle Filter

When baffle filters become coated with a grease film, they must be removed and washed. Washing depends on the construction of the filter. The filter may be simple back-washing with hot water, washed in a commercial dishwasher, or immersed in a detergent-laden solution in a wash tank. Regardless of the method of washing, the filter should be followed by draining and drying before being put back into service.

Note to Chapter 3

1. 2003 ASHRAE Applications Handbook., p. 31.3 Kitchen Ventilation.

4

Extended Surface Filters

Definition

Extended surface filters are those filters in which the filter media area is greater than the face area, or to phrase it another way, filters in which the media velocity is less than the face velocity. Face velocity is the speed of the air as it approaches the filter face (See Figure 3.1. in Chapter 3: "*Impingement Filters*"). Media velocity is the speed of the air as it moves through the filter media. The most common method of increasing media area is to pleat the filter media. Other configurations are also used and they will be discussed in this chapter.

Filters designed with extended media surface are desirable for the following reasons:

1. Media may perform more efficiently at lower media velocity. This is especially true of filters that depend on interception and diffusion principles.

2. Media resistance may be too high when the material is used in a flat sheet. This is especially true of high-efficiency media. The resistance of a filter media is, as a general rule, directly proportional to the velocity of air through it, i.e., the higher the velocity, the higher the pressure drop. Increasing media area will reduce resistance to flow until the point where media crowding - *point where pleats are packed too tightly and touch other pleats* - begins to occur within the filter.

3. Filter life will be extended. The more filter media available to hold contaminants, the longer the filter service life, all other things being held constant.

Types of Extended Surface Filters

Extended surface filters are characterized by the type of media - synthetic lofted or mat, or microglass lofted or wet-laid - and the configuration into which it is formed:

- Pleated panel – can have a cotton/ polyester blended material, all synthetic material, rarely microglass
- Pocket (bag) Filters
- Rigid box type
- Cell type with separators
- Mini-pleat style

Types of Media Used in Extended Surface Filters

Lofted Synthetic Media

Synthetic fibers are being introduced in air filters in increasing applications. These were first seen in pocket filters using air laid manufactured materials, the same as in synthetic panel ring and link filters. Also called "lofted synthetic," these filters have fiber sizes in the 30 μm range (15 denier) approximately 1 inch (25.4 mm) thick and larger and MERV numbers in the 6 - 10 range.

Synthetic Fiber Mats

Spunbond and meltblown manufactured fiber mats with fibers in the size range of 15 - 20 and 3 - 10 μm respectively, achieving higher efficiencies in the MERV 8 - MERV 15 range, are now incorporated in extended surface filters of all types. These mats, made by a variety of proprietary materials and processes, have much greater physical strength than synthetic lofted media. This allows the pockets to extend vertically without support wire or rods. Because the material is synthetic, this material may have thermally bonded (heat sealed) pockets. Synthetic mat extended surface filters do not have color coding corresponding to filter efficiency. Nanotechnology (a nanometer (nm) is one billionth of a meter) used in synthetic fiber manufacture is now coming into the market with fiber diameters in the 20 - 200 nm range (0.0002 - 0.02 μm) and, when combined with existing fibers, raises the efficiency of filter media.

Today, media materials are not filter-configuration specific, i.e. all cell-type filters use paper type medias, as in the past, and most all extended surface filters are now available with either microglass, synthetic, or some other type media. Examples in this chapter are shown with a type of media for illustration purposes only.

Lofted Microglass Media

Also known as glass mat, this media is composed of very fine glass microfibers. For example, a MERV 15 mat is composed of fibers approximately .97 μm in diameter. The fibers are formed by passing primary glass filaments in front of a gas burner that melts and blows them into fine fibers by a process known as Flame Attenuation. The fibers are sprayed with a colored binder and collected on a conveyor in the form of a mat.

The binder gives the mat physical strength and the color identifies the material. Because of their low tensile strength, and despite the addition of binders, glass mats cannot be used by themselves as air filter media. They require a backing material to provide physical support. Backing can be applied at the time the mat is formed or the mat and backer can be combined just before incorporation into a filter. The backing mat can be a synthetic material such as non-woven polyester or nylon or it may be glass scrim.

Glass mats are available in nominal thicknesses of 0.5 in. (12.7 mm) or 0.25 in. (6.4 mm). When two mats of the same material are compared, the efficiency of both thicknesses will be essentially the same. However, the thinner mat will have a lower initial pressure drop with possibly less life in service. [1]

Air filter manufacturers can order a factory run of filter mat to meet their specifications, including color. However, the principal U.S. producer of glass mat filter media identified its standard filter mats as follows:

Color	Efficiency *	Diameter **
Yellow	MERV 14-15	0.97
Pink	MERV 13	1.25
Orange	MERV 11	2.6
Tan	MERV 10	4.0

**Efficiency (ASHRAE Minimum Efficiency Reporting Value*

*** Typical Fiber Diameter,* μm

Wet Laid Microglass Media

Manufacture of wet laid media for medium and high efficiency filters is proprietary with all manufacturers. This media is made similar to cellulose paper except using glass fibers of various diameters depending on desired efficiency. Glass fibers are blended in a slurry until separated and suspended. Binders are added and the liquid fibers are allowed to flow onto a moving screen where the water is removed. The media is then moved through rollers and ovens where the material is dried.

Types of Typical Extended Surface Filters

Pleated Filters

One of the more popular configurations for extended surface filters is the pleated filter (Figure 4.1). While it may have depths of as much as 6 in. (152 mm) or as little as 1 in. (25.4 mm), the most popular size is 2 in. (50.8 mm) deep. Pleated filters are constructed using cotton/polyester blends, synthetic medias, and electrostatic (electret) medias. Some of these medias are supported by an expanded-metal or a welded-wire grid while others require no metal support.

Filters of this style usually have a higher efficiency than flat panel filters while maintaining acceptable pressure drops. In addition, pleated filters tend to have a longer service life in typical HVAC applications.

Lofted Synthetic Filters

A filter which is normally rectangular in construction is called a cube filter. Internal support grids may be sewn into the air intake (open side) of the cubes and/or metal headers utilized in order to allow for ease of use in standard HVAC filter housings and filter tracks. Another design creates several pleats that are held in position by wirework backing.

Photo Courtesy Airguard Industries

Figure 4.1 Pleated Panel Filters

Photo Courtesy Fiberbond Corporation

Figure 4.2. Self-Sealing Cube Type Filter

Most cube filters use synthetic media which is approximately 1 in. (25.4 mm) thick. One version of the pocket style uses two plies of media of different fiber size and density. The more open larger-fibered mat is on the upstream side and is intended to capture larger particles. Others use single plies whose fiber diameter depends on the manufacturer. (Figure 4.2).

The use of a factory-applied adhesive on completed cartridges for filters of this type has the advantage of reducing the tendency of the dust particles to migrate through the media. The bond between the dry media and large dust particles is relatively weak, and the dust may easily migrate without the additional retention offered by the adhesive.

Non-supported Pocket (Bag) Filters

The non-supported pocket filter is one of the more popular design high-efficiency air filters used today. Individual pockets are either stitched, heat sealed or sonic welded together, allowing for the most effective spacing of the media but preventing the pockets from "ballooning." These pockets are then fastened into a frame, usually having a minimum of six pockets and a maximum of twelve. When air flows through the pockets, they inflate, exposing all of the media to the air stream.

A filter required to meet a specific need can usually be found by proper selection of media type, length, and number of filter pockets. (Figure 4.3.)

Use in a Variable Volume System

Since the pockets of a non-supported filter are maintained in their full open position by the airflow through them, there is some concern about their use in a variable volume system. The system airflow can be significantly reduced, causing the pockets to sag, resulting in poor utilization of filter media. This issue can be addressed in a number of ways. One answer has been to develop a pocket support device which the ends of the pockets are individually attached so that they are always held vertical. Another method has been to shorten the pockets so that sagging is not significant. To compensate for this shortening the number of pockets is increased.

Photo Courtesy Purolator Products Air Filtration Co.

Figure 4.3. Non-Supported Glass Mat Pocket Filter

Some manufacturers use a stiff media or backing mat in the pockets of the filters to enable the pockets to maintain their fully open form even when there is no air flowing through them (Figure 4.4).

Photo Courtesy of Freudenberg Nonwovens

Figure 4.4. Non-Supported Pocket Filters

Rigid Box Filters

Where in-line space is restricted, rigid box deep-pleated filters are available in depths of 4 in. (102 mm), 6 in. (152 mm) and 12 in. (305 mm). One size with a nominal face dimension of 24 in. x 24 in. (610 mm x 610 mm) and a depth of 12 in. (305 mm) is rated at 2000 cfm (0.944m^3/s) and has 58 ft^2 (5.39 m^2) of media area. It is available in MERVs of 10 up to 14.

Calculating Filter Media Area in a Pocket Filter

sides of pocket	=	2
number of pockets	=	8
pocket width	=	25″
pocket depth	=	22″

Gross media area =

$$\frac{2 \times (8 \times 25 \times 22)}{144} = 61.1 \text{ square feet}$$

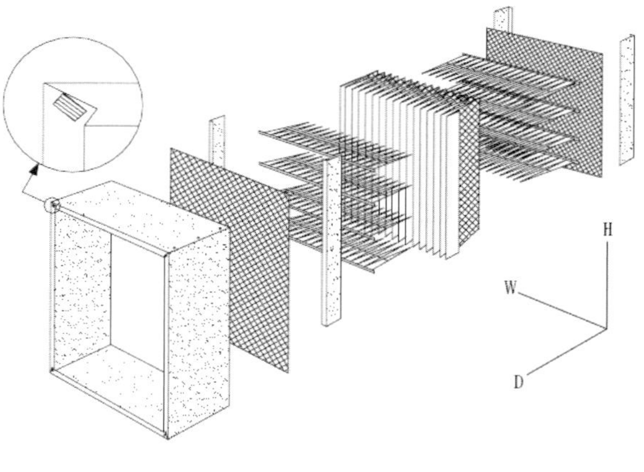

Courtesy Quality Filters, Inc.

Figure 4.5. Exploded View of Rigid Box Filter Incorporating Lofted Microglass

These filters (Figure 4.5) consist of high-efficiency media, either lofted microglass or meltblown synthetic bonded to a welded wire or expanded metal grid. The support is pleated and the pleat spacing maintained by cardboard, plastic, or metal contour stabilizers, called "fingers" on the upstream and downstream side of the pleats. The pack is sealed into a galvanized steel filter frame with diagonal supports. The sealant is frequently compressed microglass combined with adhesives of various types as needed. The final assembly is a rigid box. Rigid box filters and rigid cell filters can be manufactured in a square box configuration, or can be made with a metal "header' like a pocket filter to install in a 1 in. (25 mm) track.

This box style has the advantage of easy installation, especially in factory-fabricated air handling and rooftop units where in-line space is restricted. For any given type of filter media, rigid box filters are likely to operate at higher media velocities than other designs of extended surface filters. For this reason, they may have shorter service life. Another disadvantage to this style box filter is the higher initial resistance to air flow. This additional restriction may result in higher energy costs.

Rigid Cell Filters

Rigid cell filters are comprised of pleated wet-laid microglass with corrugated separators (Figure 4.6). They are fabricated by pleating the filter media and inserting it over corrugated aluminum separators to form a filter pack. This pack is usually sealed into a light-gage metal or wood frame, using various types of adhesives as needed. Depending on the application, the face velocity of the filter can range from 250 to 500 fpm (1.27 to 2.54 m/sec). Filters using this configuration may occupy less in-line space than alternative designs (pocket filters) of comparable capacity.

Photo Courtesy Koch Filter Corporation

Figures 4.6. Rigid Cell Filter Using Wet-laid Microglass and Corrugated Separators

Mini-Pleat Filters

Mini-pleat filters are formed by pleating wet-laid microglass or synthetic media into a filter pack using string, media ribbons, adhesive beads, or other methods to maintain pleat spacing. This pack, which may be as shallow as 1 in. (25.4 mm) deep, is sealed into a frame to form rigid panels. Panels of this construction may be configured into a "V" to give the maximum amount of filter media. Because of the greater amount of filter media incorporated into filters of this type, they have a very low pressure drop and corresponding longer filter life.

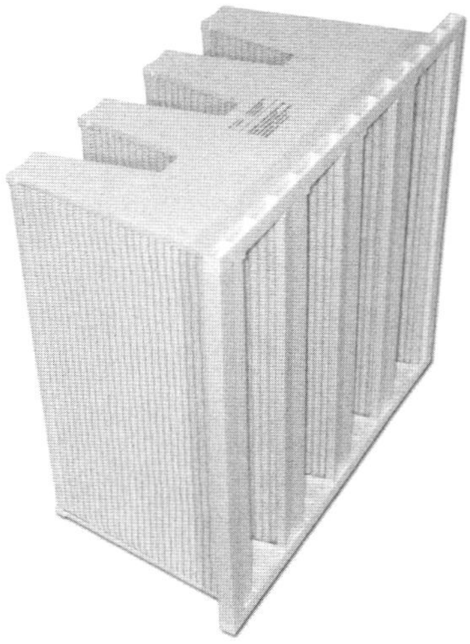

Photo Courtesy Aeolus Filter Corporation

Figure 4.7. Mini-Pleat Filter

Other mini-pleat filter designs may be 2" and 4" packs in a die cut polypropylene or metal frame.

Filter Installations

It is a basic rule that the higher the efficiency of the air filtration system, the more important it is to have an effective seal to prevent any contaminated air from by-passing the filter. For this reason holding frames for high-efficiency filters should be properly gasketed and the clamps holding the filters in place should also provide a tight seal between the filter face and the gasket. In addition, leakage between adjacent holding frames and between banks of holding frames and adjacent walls should be sealed tight by caulking or other means. Filter banks also need to be suitably reinforced to provide rigidity during system operation.

Photo Courtesy Filtration Group, Inc.

Figure 4.8. Side Servicing Housing for Extended Surface Filters

Side-access housings supplied by air filter manufacturers for air handling units are designed with channels into which the filters slide (Figure 4.8). Gasketing on the inside of the channels allows the filter frames to slide easily while preventing leakage at the top or bottom. Filters intended for side servicing should be fitted with a gasket on at least one vertical side of the filter frame so that there is a tight seal when they are butted together. The housings themselves are designed so that there is a tight fit between the filters and the access door when the proper number and size filters are installed.

Filter Performance

Typical performance data for the more popular style extended surface filters is shown in Figure 4.8. Note that these are typical data. For information about a specific product one should contact the manufacturer who can supply this filter performance information in the form of an ANSI/ASHRAE Standard 52.2 test report.

Deep-Pleated Residential Filters

This type of residential/small commercial air filter utilizing pleated media is generally available in many sizes and is used on residential systems. Media is similar to that used in pleated filters – typically MERV 6 to 11. See Figure 4.9.

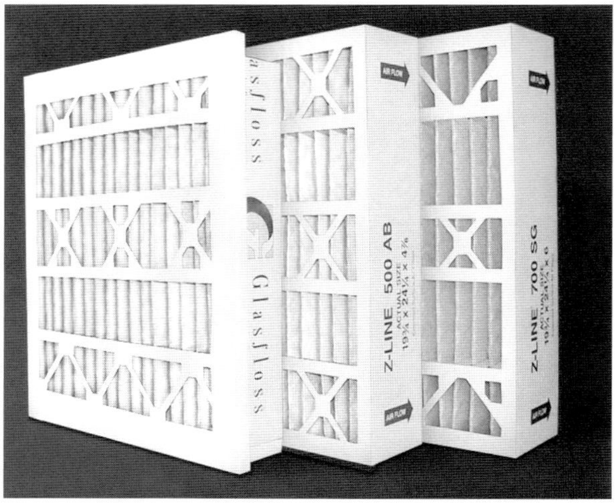

Photo Courtesy Glasfloss Industries, Inc.

Figure 4.9. Deep-Pleated Residential Filters

Note to Chapter 4

1. Manville Publication "Manville Fiber Glass Filtration Media" FGM-51 (8-89) (Fiber diameter information supplied in a separate communication.)
2. Kimberly Clark.

Figure 4.8. - Typical Performance Values of Extended Surface Filters Face Dimensions 24 in. x 24 in. (610 mm x 610 mm).
Rated Capacity 2000 cfm (0.945 m³/s)

Type Filter	Media Type	Depth inches (mm)	Average Efficiency (1)	ANSI/ASHRAE Standard 52.2 MERV	Initial Pressure Drop in. w.g. (Pa)	Media Velocity fpm (m/s)	Media Area ft² (m²)
Non-Supported (Bag)	Glass or Synthetic	29 (737)	90 - 95 80 - 85 60 - 65 45 - 50 40 - 45	14 13 11 9	.50 - .60 (124 - 149) .40 - .50 (99.5 - 124) .35 - .40 (87.1 - 99.5) .15 - .20 (37.3 - .49.8)	24 (.122)	84 (7.8) (85)
Non-Supported (Cube)	Highloft Polyester	15 (381)	30 - 35	8	.25 - .30 (62.2 - 74.6)	63 (.102)	32 (2.97)
Box (Corrugated Separators) Rigid Cell	Microglass	12 (305)	90 - 95 80 - 85 60 - 65	14 13 11	.65 - .75 (162 - 186.8) .50 - .60 (124 - 149) .40 - .50 (99.5 - 124)	16 (.081)	125 (11.6)
Rigid Box	Glass (highloft) or Synthetic	12 (305)	90 - 95 80 - 85 60 - 65	14 13 11	.65 - .75 (162 - 186.8) .50 - .60 (124 - 149) .60 - .65 (149 - 162)	34 (.088)	58 (5.39)
Pleated Panel	Polyester / Cotton or Synthetic	2 (50.8) 4 (101.6)	20 - 30 20 - 30	6 - 8 6 - 8	.20 - .35 (49.8 - 87.1) .20 - .35 (49.8 - 87.1)	111 - 166 83 (.422)	12 - 18 24 (2.23)
	Glass	2 (50.8) 4 (101.6)	30 - 40 30 - 40	7 - 8 8 - 11 7 - 8 8 - 11	.20 - .35 (49.8 - 87.1) .30 - .40 (74.9 - 99.6) .20 - .35 (49.8 - 87.1)	125 (.635) 83 (.422)	16 (1.49) 24 (2.23)
Mini-pleat V-Style	Microglass	12 (305)	90 - 95 80 - 85 60 - 65	14 13 11	.35 - .40 (87.1 - 99.6) .25 - .35 (62.2 - 87.1)	10.5 (.053)	190 (17.7)

Note: Efficiency is ANSI/ASHRAE Standard 52.1 atmospheric dust spot discoloration efficiency (percent)

5

HEPA, ULPA and Super ULPA Filters

Definition

The Institute of Environmental Sciences and Technology (IEST) defines a HEPA filter as:

> An extended-medium dry-type filter in a rigid frame, when tested at rated airflow having minimum particle-collection efficiency of 99.97% for 0.3 micrometer (µm) mass median diameter particles of DOP when tested in accordance with MIL-STD-282, particles of thermally-generated DOP particles or specified alternative aerosol.[1]

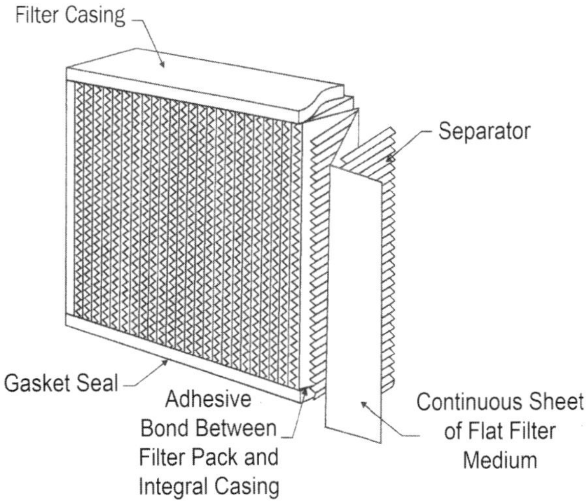

Figure 5.1. Components of HEPA Filters

General Description of Filter

The general components of HEPA filters are identified in Figure 5.1. This drawing should be considered generic because, as will be noted later, while not all these components may be found in each filter, there is an equivalent function for each.

Media is the filtering material. It is a paper composed of microglass fibers, synthetic fibers, expanded film such as polytetrafluoroethylene (PTFE), or other types of fibers, that can be pleated back and forth to form a compact filter element. Close-pleating is necessary to fit all the required media into the desired space because the paper has a high resistance to airflow and the media velocity is usually in the range of 6 fpm (.03 m/s).

Separators are devices which support paper media pleats. Separators also provide channels through which the air can flow to reach the media in a more laminar flow pattern and then, after passing through the media, flow out of the filter.

Filter pack is the combination of media and separators.

Filter frame (cell side) is the rigid box into which the pack fits.

Sealant is the adhesive or other material used to create a leak-proof seal between the filter pack and the filter frame. It may also describe the material used to splice or repair a leak in the filter media.

Filter Gasket Seal prevents any bypass of unfiltered air around the filter where it bonds to the holding frame. In most instances, it is either a closed-cell neoprene gasket attached to the face of the filter frame, or a groove in the frame to allow a knife-edge to penetrate a non-Newtonian gel.

Filter Media

The original asbestos-cellulose paper provided an efficiency of 99.95% in removing 0.3 μm particles by the DOP Test method (See Chapter 8: "*HEPA and ULPA Filter Testing*"). It was succeeded by an all-glass thick filter paper which had the same general performance characteristics. This, in turn, was followed by a thin all-glass paper which had an improved DOP efficiency of 99.97%.

Figure 5.2. Temperature and Humidity Limitations of HEPA Filter Components

Component	Material	Max Temperature °F (°C)	Max. R.H. Percent
Frame	Exterior-Grade Plywood	250 (121)	100
	Fire-Retardant Plywood	250 (121)	100
	Particle Board	250 (121)	80
	Aluminum	750 (398.9)	100
	Galvanneal Steel	750 (398.9)	100
	304 Stainless Steel	2000 (1093.3)	100
Media	All-Glass Waterproof	750 (398.9)*	100
	Mil Spec (MIL-F-51079) ULPA	750 (398.9)*	100
	Acid Resistant	750 (398.9)*	100
	PTFE Membrane	450 (232)	100
		545 (285)***	100
Separators	Paper (Special Treated)	250 (121)	100
	Aluminum	750 (398.9)	100
	Vinyl-coated Aluminum	400 (204)	100
	Stainless Steel	2000 (1093.3)	100
Sealant	Neoprene Base	200 (93.3)	100
	Solid Polyurethane Ceramic	250 (121)	100
	RTV Silicone	2000 (1093.3)	100
		500 (260)	100
Gaskets	Closed Cell Neoprene Silicone	200 (93.3)	100
	Rubber	500 (260)	100
	Silicone Fluid Seal	500 (260)	100
	200 Compressed Glass**	750 (398.9)	100

Notes:
** Material remains functional but loss of binder reduces strength,*
*** Loss of compressive strength occurs over 250°F (121°C)*
**** Wikol M. et. al. Expanded Polytetrafluoroethylenes (PTFE) and their Application, W.L. Gore and Associates, Elkton, MD*

A glass-asbestos paper was also offered. The inclusion of asbestos into the media improved its resistance to acid gases, especially hydrofluoric acid, and was used extensively at government nuclear facilities. As concern developed about even small amounts of asbestos in a product, this media was abandoned.

Continued requirements for higher filter efficiency resulted in newer media carrying the generic title of ULPA (Ultra Low Penetration Air). The Institute of Environmental Sciences and Technology (IEST) defines an ULPA filter as:

> "A throwaway, extended-medium, dry-type filter in a rigid frame, having a minimum particle collection efficiency of 99.999% (that is, a maximum particle penetration of 0.001%) when tested in accordance with the methods of IEST-RP-CC007."

Note: Super ULPA (Super Ultra Low Penetration Air) is a name for filters offered by certain manufacturers, having an efficiency of 99.9999%, and designated as a Type G filter by IEST-RP-CC001.4.

The formulation of filter media by different paper manufacturers is proprietary. However, paper media is made by beating fibers (usually all glass) under controlled conditions in a properly buffered water base until the fibers are all separated and suspended. Proprietary binders are added, usually modacrylic suspensions of different types. Antifungal inhibitors may also be used. The suspension of glass fibers is allowed to flow onto a continuous screen (Fourdrinier screen) where it is dewatered as it moves along. At the end of the screen line, it is moved onto drying rolls where the balance of the water is removed. Before rolling, the media may be waterproofed, usually by treatment with a silicone-based material.

Another category of filter media is a membrane material made of expanded PTFE (polytetraflu oroethylene). It is has the advantage of being a non-shedding organic material. Its non-shedding characteristics make it desirable for any super-clean installations and its organic base means that it is free of any elements such as boron that can, if off-gassed, be a contaminant in the manufacturing of certain microelectronics. The material can be formulated for a variety of efficiencies. To date the principal application for this type of filter is in the filter category referred to as ULPA filters.

Separators

Originally, corrugated separators were made of kraft paper. Kraft paper was replaced by aluminum foil which is the most popular material for this application today.

Other separator materials include stainless steel and coated aluminum. Polyvinyl chloride and polystyrene have also been used in the past.

Stainless steel foil is very sharp and when used as separators can easily damage filter media. To prevent this, metal separator material is hemmed to produce rounded edges. Hemming has become popular for use with aluminum foil corrugations to provide greater physical strength and better resistance to unsightly damage due to exposed corrugation edges.

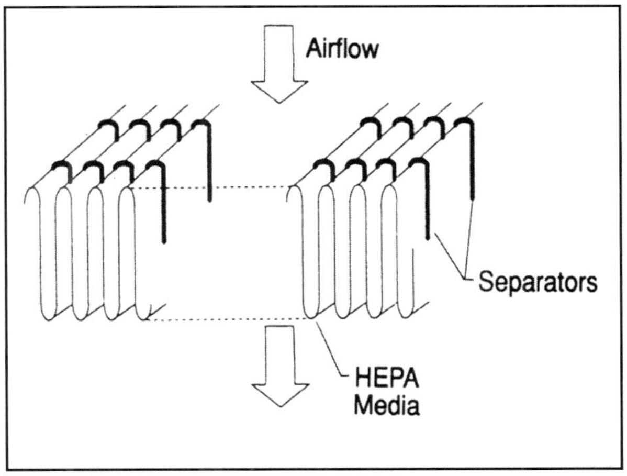

Figure 5.3. Close-pleated Filter Pack with String Separators

There is a design of HEPA filters in which a 12 in. (305 mm) deep filter element, instead of consisting of one deep-pleated filter pack, is composed of smaller [(1.08 in.) (2.75 mm)] deep filter elements in a "V" configuration inside a filter frame. These shallow filter elements do not use corrugated separators. Instead, they accomplish pleat spacing by attaching correctly spaced narrow ribbons of media, string, or similar material to the media before it is pleated. The ribbons or other materials butt against each other as the media is pleated, allowing space so that the air can flow through the media and then out the other side (Figure 5.3).

Another method of fixing pleat spacing is by the use of thin beads of hot melt adhesive on both faces to control pleat shift. Hot glue is also used as a separator where beads are placed on corresponding areas of the media so that, when pleated, they meet and form a space between pleats.

Filters are available where the media is formed with built-in embossed grooves. When the media is pleated, the grooves then butt against one another creating channels through which the air can flow. The elimination of corrugated separators reduces the resistance of the filter pack. However, it also makes the pack more flexible. To compensate for this, reinforcing struts are inserted inside the pack as needed.

Mini-pleat

Thinner (mini-pleat) filter elements are used in many cleanroom downflow air supply modules (see Figure 5.4) where the filter face velocity is 100 fpm (0.508 m/sec). Under this operating condition, the pressure drop of a mini-pleated filter element is comparable to a 6 in. (152 mm) deep HEPA filter with corrugated separators. Because of the close-pleating of the shallower unit, both filters have a comparable amount of filtering media and will have corresponding life cycles. The mini-pleat filter is frequently used where overhead space is limited. Units are available in depths of as little as 1 in (2.54 cm), however usually this short of a depth has a lower air flow rating or possible higher resistance to airflow.

Figure 5.4. Self-Contained Terminal Ceiling Module

In addition to string and ribbon separators, mini-pleat filters are available in which dimpling of the media maintains pleat spacing.

Sealants

A variety of materials are used to seal a filter pack into a filter frame. The earliest materials were rubber-based and then neoprene-based adhesives, that were forcefully sprayed into the ends of the pleats to seal them and prevent any leakage. A corresponding coating was sprayed on the filter frame so that when the whole frame was assembled, the two adhesive coatings combined, blocking off any leak paths. This process was not 100% successful because of the viscosity of the adhesive and its tendency to skin-over very quickly. Another method used a polyurethane material that was foamed in place. Other materials have been trowel-applied and have flow characteristics that allow them to penetrate the pleats and accomplish a seal.

One design requires a metal frame and uses two metal pans as the top and bottom of the filter frame. One pan is partially filled with a two-part polyurethane mixture which is liquid when first formed but sets up as a hard seal when fully cured. One of the ends of the assembly of the filter pack and the two side panels are immersed in this mixture and properly held in a fixture until the sealant sets. The operation is then repeated with the opposite end. This method provides a consistent leak-proof seal.

For high temperature applications special adhesives must be used. Currently many high-temperature filters are made with a thin layer of ceramic cement troweled into the pack to seal it. A glass mat is applied over the ceramic material and compressed during installation to prevent leakage between the pack and the filter frame. Filters with this construction are rated at 99.97%

DOP efficiency. Because the compressed glass mat only friction-bonds the pack and frame, the filter is supplied with a downstream face bar to prevent pack movement during use.

Frames (Cell Sides)

The original framing material for HEPA filters was exterior grade plywood. This World War II wartime usage continued for economic reasons. For high temperature and moisture conditions, cadmium-plated steel was used. Exterior grade particle-board was used as an economic replacement for plywood. Where fire resistance was required and metal was not allowable, fire-retardant exterior grade plywood was used. The fire-retardant treatment was performed after the plywood had been manufactured by treating it with metallic salt solutions. Fire retardant particle-board was available in interior grade but not exterior grade. The regular exterior grade material could not be treated to be made fire-retardant.

Cadmium-plated steel is no longer available. Alternative materials include galvanized and galvanneal steel, aluminum, and stainless steel. Chromized steel was used for high-temperature applications. It has now been replaced by muffler-grade steel. The choice usually depends on the chemical application and temperature involved.

Filter Performance

Filter performance testing does not measure efficiency, it measures penetration, i.e., the fraction of a challenge aerosol which passes through the filter. Each filter is individually tested. This is absolutely critical. Using a simple description, a challenge aerosol is applied to the upstream side of a filter. This challenge aerosol concentration is measured and set as a base line. A sample is taken on the downstream side of the filter and referenced to the upstream sample. Efficiency is determined by subtracting the penetration (percent) from 100%. The original maximum penetration was 0.05% so that the efficiency was 99.95% (100 minus 0.05).

Later improvements in media manufacture and materials allowed efficiency to be extended to 99.97% without any increase in pressure drop. It should be noted that these efficiencies are minimum values and that actual penetration test measurements usually show efficiencies much higher than this. Now newer media are available with much higher efficiencies and somewhat higher pressure drops. Some of these efficiencies are now so high that traditional testing methods are no longer the only method of measuring performance. These testing methods are discussed in Chapter 8: "*HEPA and ULPA Filter Testing.*"

For every filter operating at a specific cfm rating, there is a most-penetrating particle size (MPPS). This is the size particle on which the filter has the highest penetration and lowest efficiency, i.e. the worst-case particle size with respect to filtration efficiency. Efficiency will be higher on particles both larger and smaller than this MPPS. This is not a significant concern for typical ventilation filters, but is important wherever high collection efficiency on small particles is required, such as in the case of HEPA filters. Particles 0.3 μm in size were originally chosen as the test challenge because it had been calculated that this would be the most-penetrating particle size for HEPA filters.[2]

The need for better filtration and means of verifying filter performance has lead to the development of newer ways of testing HEPA and ULPA filters. These requirements have primarily grown from applications in the microelectronics industry. Figure 5.7 from IEST-RP-CC001.4 summarizes the types of filters now available, the methods of testing, and typical applications.

When the laminar flow concept was introduced for cleanrooms, the penetration location in HEPA filters became a matter of concern. Wherever there was leakage, a stream of unfiltered air could pass through these leakage openings. Because the flow was "laminar" (non turbulent) the stream of dirty air remained intact, potentially contaminating a product in the "clean" air stream. This resulted in the requirement that a HEPA filter for use in a laminar flow application had to be leak-free. When leak-free filters were made, their minimum efficiency was typically 99.99%. Penetration comes either from "pinhole" leaks too small to detect or from the limited capability of the filter media. (Leak testing is covered more completely in Chapter 8: "HEPA and ULPA Filter Testing."

Figure 5.7. Categories of HEPA and ULPA Filters[4]
IEST-RP-CC001.4

Filter Type	Description	Application
A	99.97% efficient on 0.3 µm MMD particles at rated flow when tested per MIL-STD-282	Commercial, industrial, general HVAC applications. Note: Some of these applications may not allow for the filters to be tested with an oil-based challenge and should use a Type H filter
B	99.97% efficient on 0.3 µm MMD particles at rated flow and 20% of rated flow when tested per MIL-STD-282	Nuclear, Department of Energy, Department of Defense and other applications requiring the filters meet ASME AG-1, Section FC
C	99.99% efficient on 0.3 µm MMD particles at rated flow when tested per MIL-STD-282 and has been leak tested (scan tested) per IEST-RP-CC034	Pharmaceutical manufacturers, biotechnology, semiconductor, hospitals and other cleanroom applications. Note: Some of these applications may not allow for the filters to be tested with an oil-based challenge and should use a Type K filter
D	99.999% efficient on 0.3 µm MMD particles at rated flow when tested per MIL-STD-282 and has been leak tested (scan tested) per IEST-RP-CC034	Pharmaceutical manufacturers, biotechnology, semiconductor and other cleanroom applications. Note: Some of these applications may not allow for the filters to be tested with an oil-based challenge and should use a Type J filter
E	Designed, tested, and constructed in accordance with MIL-F-51477 or ASME AG-1, Section FC	Animal disease laboratories
F	99.999% efficient when tested at rated flow per IEST-RP-CC007, determined as the lower efficiency at 0.1-0.2 µm or 0.2-0.3 µm	Pharmaceutical manufacturers, biotechnology, semiconductor, aerospace and other cleanroom applications
G	99.9999% efficient on MPPS when tested at rated flow per IEST-RP-CC007 and has been leak tested (scan tested) per IEST-RP-CC034	Semiconductor, aerospace and other applications requiring < ISO Class 4 cleanrooms per ISO 14644-1
H	99.97% efficient when tested at rated flow per IEST-RP-CC007, determined as the lower efficiency at 0.1-0.2 µm or 0.2-0.3 µm	Commercial, industrial, general HVAC applications. Note: This filter should be used when the customer's filter size or rated flow may exceed the manufacturer's MIL-STD-282 test equipment capabilities
I	99.97% efficient when tested at rated flow per IEST-RP-CC007, determined as the lower efficiency at 0.1-0.2 µm or 0.2-0.3 µm. It has also been leak tested at 20% of rated flow per IEST-RP-CC007	Nuclear, Department of Energy, Department of Defense and other applications requiring the filters meet ASME AG-1, Section FC
J	99.99% efficient when tested at rated flow per IEST-RP-CC007, determined as the lower efficiency at 0.1-0.2 µm or 0.2-0.3 µm. It has also been leak tested (scan tested) per IEST-RP-CC034	Pharmaceutical manufacturers, biotechnology, semiconductor, hospitals and other cleanroom applications. Note: This filter should be used when the customer's filter size or rated flow may exceed the manufacturer's MIL-STD-282 test equipment capabilities
K	99.995% efficient when tested at rated flow per IEST-RP-CC007, determined as the lower efficiency at 0.1-0.2 µm or 0.2-0.3 µm. It has also been leak tested (scan tested) per IEST-RP-CC034	High purity pharmaceutical manufacturers, biotechnology, semiconductor. Note: This filter should be used when the filter size or rated flow may exceed the manufacturer's MIL-STD-282 test equipment capabilities

Holding Devices for HEPA Filters

Individual Holding Frames. The earliest holding devices for HEPA filters were box frames that could form a filter bank by being set next to or above one another in the form of building blocks. Caulking was applied to the sides of the frames before the bank was assembled and the frames riveted together.

The simplest design is a holding frame into which the filter can fit (Figure 5.6). The filter is supplied with gaskets on the edges of one side. When the filter with the gasket side first is placed into the holding frame, it is held in place by clamps which exert pressure to compress the gasket and make the seal between the frame and the filter.

Some clamping devices are offered with spring-loading mechanisms, that when fully compressed, exert the correct amount of pressure against the gasket.

HEPA filters with closed-cell neoprene gasketing must be completely and correctly compressed into the HEPA filter holding frame. Because surface plane tolerances vary, experience shows that gasketing must be compressed at least 80% in order to affect a seal. This equates to 20 pounds per square inch of gasket area or a total of 1400 pounds for a typical 24 in. x 24 in. HEPA filter.

Figure 5.6. Typical Holding Frame for HEPA Filter

It is recommended that the gasket be coated with a silicone gasket release material to prevent the gasket from becoming a problem upon replacement. It is also recommended procedure to compress the gasket 50% for the first two weeks, and then re-torque to 80% compression. While holding frames for lower efficiency filters usually have permanent gaskets attached to them, the holding frames for HEPA filters are plain and the filters are gasketed. This assures that a good seal can be provided with each filter change.

Another holding device utilizes a fluid seal. A fluid seal utilizes a non-Newtonian fluid (gel like) filling a groove in the sealing side of the filter frame. When the filter is pushed into the holding frame, a knife edge penetrates the non-Newtonian fluid and creates a seal. This method of seal is especially popular where the filter is installed in a ceiling grid and the weight of the filter holds it against the knife edge. Spring loading is unnecessary. Fluid seals are self-repairing and do not adhere to the knife edge when the filter is withdrawn. Fluid seal filters typically make sealing the filters to a framing system simpler and easier. As with all seals, compatibility of materials must be evaluated.

Room Side Replaceable Wall and ceiling framing modules were developed for clean space applications. These are usually made from extruded aluminum and welded into assemblies which can be hung from ceilings or form walls of cleanrooms. Many of these designs are proprietary. The modules have the advantage of being available in larger sizes than the normal 24 in. x 24 in. (610 mm x 610 mm) face dimensions of a holding frame. Also, since the modules are all-welded construction, they present less chance of leakage than riveted assemblies of holding frames. See Figure 5.8. Special heavy-duty modules were also developed for nuclear applications of HEPA filters. These are described in a DOE (AEC) publication.[4]

Disposable HEPA Filter Modules are designed so that the filter pack is incorporated into the center of a housing with extended flanges. End covers, incorporating correctly sized fittings, are attached to the housings after all joints have been caulked. When the filter becomes loaded, the whole assembly is discarded. This concept is used in a variety of configurations.

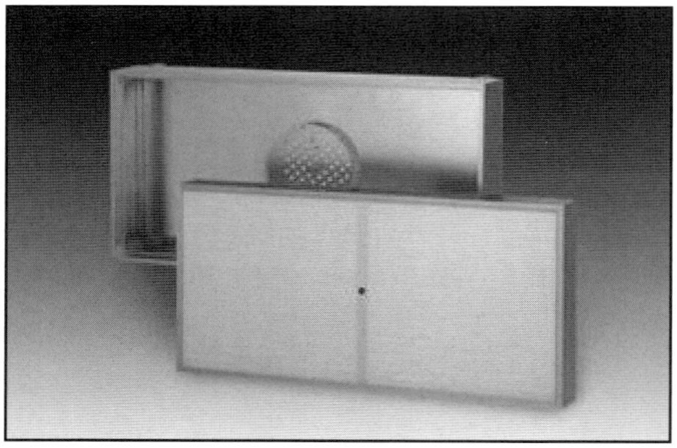

Figure 5.8. Room-side Replaceable Terminal Filter

Side-servicing housings are used where in-line space is at a premium. In this design HEPA filters are slid into a housing from one or both sides. Care must to be taken to make certain that the filters line up with the matching flanged openings in the housing. When this is done, clamping mechanisms that vary with different manufacturers, compress the filter gasket to create a seal between the filter and the housing. Corresponding housings using gel seals are also available. (Figure 5.10.)

Photo Courtesy Filtration Group, Inc.

Figure 5.10. HEPA Housing for Side-Service

Bag-in, bag-out housings are modifications of side-service housings, (Figure 5.11) intended for those situations where hazardous material, radioactive or toxic dusts are handled. They are designed so that a contaminated filter is removed in a sealed plastic bag and a clean replacement filter installed without breaking the seal between the housing and the surrounding environment. The steps involved in this process are shown on the following page in Figure 5.9. *It is important to point out that the actual bag-in, bag-out device and the operations involving its use are more hazardous and complex than the schematic drawing details.* Figure 5.9 is for illustration purposes only and should not be construed to represent a training protocol for this hazardous procedure. **Contact NAFA for details on Certification for this protocol.**

Photo Courtesy P & G Manufacturing

Figure 5.11. Bag-in Bag-out

HEPA "Like" Filters – 95% DOP

The suitability of the HEPA filter configuration for filters with efficiencies of MERV 15 or lower has been discussed in the chapter on extended surface filters. There is one category of filters sometimes referred to generically as biofilters that have been used extensively to fill the gap between HEPA filters and those tested by the ANSI/ASHRAE Standard 52.2 method. This filter has a DOP efficiency of 95% and an ASHRAE MERV of 16. The 12 in. (305 mm) deep version has a clean pressure drop of 0.5 in. w.g. (124 Pa) when operating at a face velocity of 250 fpm (1.27 m/s). When operated at 500 fpm (2.54 m/sec) face velocity, its clean pressure drop will be 1.0 in. w.g. (249 Pa), but its DOP efficiency will be less than 95%. This filter can look exactly like a HEPA filter and frequently can only be distinguished by the manufacturer's label or by markings given to the media by the media

Figure 5.9. Bag-in, Bag-out Housing for HEPA Filters

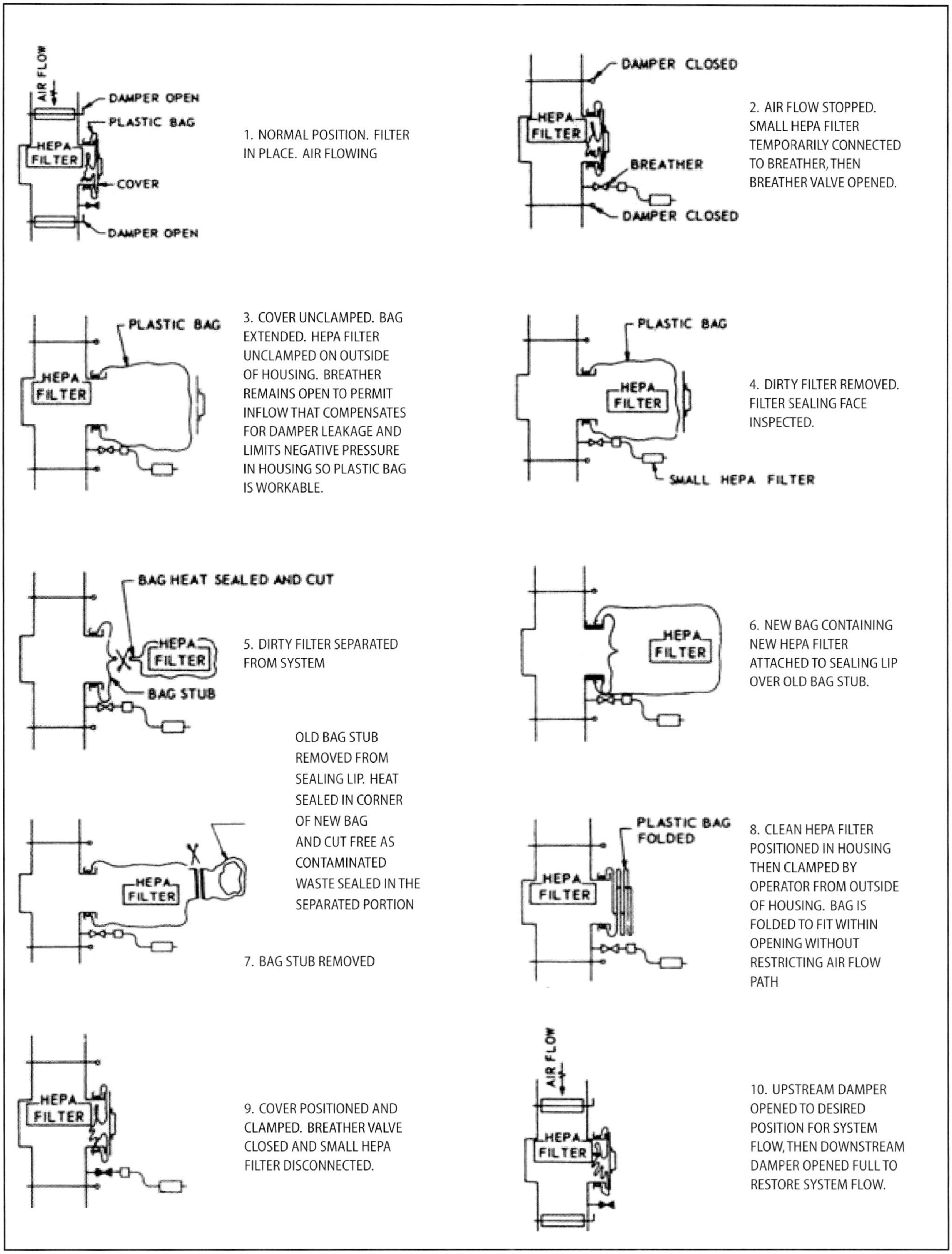

manufacturer. It is also important to note that these filters are not individually tested for penetration/efficiency like HEPA filters but have media that is manufactured as 95% DOP media.

Corresponding performance is also available in units consisting of mini-pleat V-cells. One model has a DOP Efficiency of 98%. When rated at 2000 cfm (943.8 1/s), its clean pressure drop is 0.76 in. w.g. (189 Pa).

Super ULPA Filters

As mentioned previously, a filter with a higher efficiency than an ULPA filter was offered around the late 1980's. It had a DOP efficiency of 99.9999% and the 12 in. (305 mm) deep version had a clean pressure drop of 1.1 in. w.g. (274 Pa) when operating at a face velocity of 250 fpm (1.27 m/s). This filter has helped meet the requirement for cleaner air in facilities needed for the manufacture of microelectronics. It is identified by the generic name SULPA (Super Ultra Low Penetration Air) filter.

It was previously noted, the most-penetrating particle size (MPPS) for a HEPA filter was calculated to be 0.3 µm. The development of the laser particle counter has provided evidence that the most-penetrating size particle is less than this and depends not only on the filter media but also on the velocity of air through it.

Notes to Chapter 5

1. IEST-RP-CC001.4 HEPA and ULPA Filters, Institute of Environmental Sciences and Technology, Mt. Prospect IL.
2. Langmuir, Report on Smokes and Filters OSRD 865 (1942).
3. See Note 1
4. IEST-CC001.4 HEPA and ULPA filters, ibid page 25.

6

Air Cleaners

Media Air Cleaners
***Electronic Air Cleaners** - Also known as Electrostatic Precipitators*
Agglomerators

Introduction

While every air filter could be called an "air cleaner," air cleaners as used in this chapter are defined as:

- equipment that is built for an extra cleaning, i.e. used in combination with other HVAC filtration
- built for a specific contaminant removal application
- derive their power source from A/C current. In this chapter different types of air cleaners are discussed and classified by the method of contaminant removal.

Media Air Cleaners

Media air cleaners usually are found in two configurations, either portable units or in-duct units.

Portable media air cleaners are those manufactured for cleaning specific areas at the source of the contaminant such as a room or several rooms. Portable media units come in a variety of types that include HEPA filtered, carbon filtered, UVC and PCO irradiation and/or some combination of several of these configurations.

Portable media air cleaners are convenient because they can be moved around from place to place or mounted in an area above a ceiling, are small and use standard 110 VAC current, have their own internal blower, and units and filter replacements may be purchased at many different locations. See Figure 6.1.

Photo Courtesy Honeywell Commercial Products

Figure 6.1. Portable Media Air Cleaners

Media in-duct units are those specifically manufactured to be inserted in return air grilles, in the ductwork upstream of the HVAC unit, or installed on the downstream side of the HVAC unit. Most consist of a large surface area of media to give a low pressure drop, are pleated and installed into a wide filter holding frame, and accommodate MERV 8 or higher filters. See Figure 6.2.

Photo Courtesy General Filters, Inc.

Figure 6.2. In-duct Media Unit

In-duct units are preferred when the desire is to clean all of the air going through the HVAC unit since it uses the existing fan of the HVAC unit to move the air. With a media duct mounted filter, many manufacturers recommend using the HVAC fan in the "on" position for continuous air cleaning. Filters can last anywhere from 90 days up to a year in service depending on the amount of media in the filter, the efficiency, and the amount of contaminants generated in the space.

Advantages of portable media air cleaners include ease of use and cleaning (disposal of a media filter), flexibility of different configurations of filters including particulate and molecular, and a variety of different sizes and styles.

Disadvantages include higher pressure drop (in-duct unit) than ESP's, inability to wash and reinstall the filter, possible noise factor (portable unit) and, in some units, a proprietary filter holding frame that may restrict purchase of replacement filters to one manufacturer.

Electronic Air Cleaners

Electrostatic Precipitators (ESP)

The principle of electrostatic precipitation as a means of stack gas cleaning goes back to the earliest part of the twentieth century. The process consisted of applying a high voltage (50 to 100 kV) to electrodes installed between grounded plates. The electrostatic field established by this difference imposed a charge on all the dust particles passing through the field. The particles were then attracted to oppositely charged plates where they were captured. This type of equipment could not be used in air conditioning systems because the concentration of ozone formed by the electrostatic field was irritating to the human respiratory system.

Dr. Gaylord Penny of the Carnegie Institute of Technology determined that the generation of ozone could be nearly eliminated if electrostatic precipitation was done in a two-stage operation.

In the first stage (ionizer section), the particles in the air stream were given an electrostatic charge. In the second stage (collector section), these particles were removed from the air stream by electrostatic attraction to oppositely charged plates (see Chapter 2, Figure 2.12). In the original design, a 12 kV charge was placed on the ionizer section to electrostatically charge the particles. This section was made small and was designed with ionizing wire and grounded struts so that ozone generation was minimized. In the collector section, the alternately charged plates were sized and spaced so that a 6 kV charge was enough to collect the particles but not enough to create ozone. Devices used in HVAC air conditioning and ventilating systems are usually called electronic air cleaners to distinguish them from the earlier high-voltage, stack gas cleaning electrostatic precipitators. For residential or light commercial applications, electronic air cleaners come in either portable or in-duct units and are used exactly like media air cleaners.

Photo Courtesy Second Wind Air Purifiers

Figure 6.3. Ceiling Mount Air Cleaner

Self-Contained ESP

There are a variety of self-contained electronic air cleaners available on the market today. Each is supplied within its own enclosure that includes a blower to move the air through it. A typical ceiling mounted unit is shown in Figure 6.3.

It should be noted that a device should not be called an electronic air cleaner unless the collecting plates or the medium are electrostatically charged by an external power source. There is a category of filters described as "passive electrostatic filters." (See Chapter 2). One such type uses charged filter media (electrets) to enhance efficiency. Another type relies on the air moving through the media (usually polypropylene) to create a charge. Application of Electronic Air Cleaners:

1. They should have enough air-handling capacity so that they can provide the recommended air changes in the space where they are used. A standard rule of thumb is to assure at least 10 air changes per hour (ACH) in the space to be cleaned.
2. They should be easy to clean and maintain.
3. They should be positioned in a space so that recirculating air patterns are generated within the area served. Such recirculation carries airborne contaminants to the air cleaner. This is called a coanda airflow pattern.

Efficiency

It is known that electronic air cleaners have the highest efficiency when the collector plates are clean. The ANSI/ASHRAE Standard 52. 2 filter test method does not provide for the testing of portable media or electronic air cleaners. Specifically, the loading of an electronic air cleaner with synthetic test dust that contains carbon black tends to short out the electric connections and renders the unit inoperable. These units are tested to a different test method known at the Association of Home Appliance Manufacturers (AHAM) ANSI/AHAM AC-1 (see Room Air Cleaners – Chapter 7). It is not uncommon to see a manufacturer's literature stating that electronic air cleaners are effective on particles down to 0.01 µm or smaller. Usually no information is given about the fractional efficiency of the device, that is, its efficiency on particles of a certain size. However, studies on this subject do appear in technical literature. Figure 6.4 shows the results of one series of tests measuring the fractional efficiency of a specific electronic air cleaner on different size dust particles at different airflow velocities through the unit.[1]

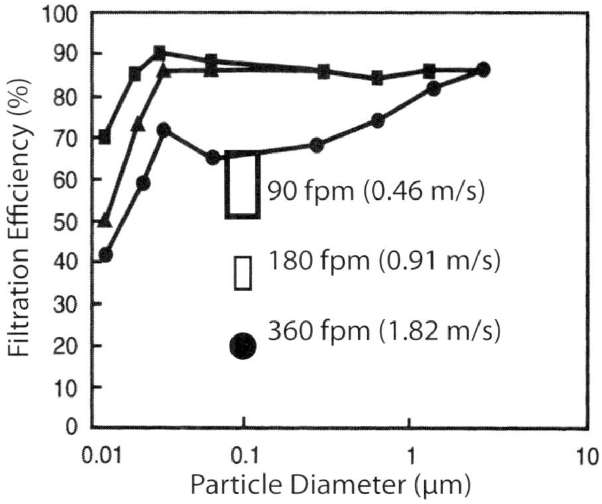

Figure 6.4. Electronic Air Cleaner Efficiency vs. Particle Size at Different Flow Rates

Factors Affecting Electronic Air Cleaner Design and Efficiency

The ability of an electronic air cleaner to remove dust from an air stream depends on a variety of factors. Some of these are:

Particle size: Very large particles in an air stream have relatively high inertia and may not be in the collecting field (between the parallel charged plates) long enough to be attracted to the plates. On the other end of the spectrum, very small particles may be so buffeted by gas molecules that attraction to an oppositely charged plate is impeded or blocked.

Velocity through the filter: Electronic air cleaner efficiency is inversely proportional to velocity through the collector section. The slower the movement of air through the collector section, the better the chance that collection will be accomplished. Electronic air cleaners require uniform airflow across the face of the filter to ensure maximum efficiency.

Voltage on ionizer and collector: The usual value for ionizer section voltage is 12 kV and for the collector section 6.0 kV. Upward variations are likely to result in ozone formation and downward variations will result in lower efficiency. Lower voltages are used in self-contained commercial and residential electronic air cleaners.

Collector plate spacing: The space between the collector plates must be optimized. If it is made smaller, the collection efficiency will improve because the distance a particle must be deflected to a collector plate is reduced. However, closer plate spacing makes maintenance cleaning of the collector section more difficult and leads to the possibility of a bridging short-circuit between oppositely charged plates. Increased plate spacing will reduce collection efficiency.

Collector plate loading: As the plates in the collector section become loaded with dust, the dust serves to insulate the charge and reduce the overall efficiency of the unit.

Ionizer spacing: It is important to make sure that particles passing through the ionizing fields are given an adequate charge. The ionizing field strength depends on the space between the ionizing wire and the oppositely charged strut. If the ionizing field is too strong, ozone formation may take place. If it is too weak, particle charging may be incomplete.

Ozone: Ozone (O_3) consists of three oxygen atoms and is a relatively unstable molecule. In high concentrations, it is actually a toxic gas. The EPA has set guidelines that establish maximum outdoor exposure levels of this gas for human health and safety. The FDA has established a limit of 50 parts per billion of ozone for medical devices. Currently, no federal agency has set specific safe limits of ozone emissions for air cleaning devices. However, UL Standard 867 has set limits on ozone generation (see page 6.7, this chapter).

Ozone is not healthy to breathe for those with respiratory diseases such as asthma and Chronic Obstructive Pulmonary Disease (COPD) or for infants or the elderly, even in very small parts per billion and is not recommended for use as an air cleaner. Not only can the ozone itself be detrimental, it can also create chemical reactions with other substances in indoor air that add to indoor air pollution. For example, researchers have found that ozone reacts with various volatile organic compounds (VOC's), especially terpenes, to create large quantities of sub-micron sized particles. Most scented products contain significant levels of terpenes. Another by-product resulting from indoor air ozone chemistry is formaldehyde - a known carcinogen. Ozone formation is minimized or eliminated by proper ionizer design and by proper maintenance of an electronic air cleaner or by the addition of an afterfilter of activated carbon.

Agglomerator Units

In large commercial or industrial applications, electronic air cleaners use dry collector plates in a two-step process known as agglomeration.

In these applications, as dust particles accumulate on the collector plate surface, the attraction between collected particles becomes

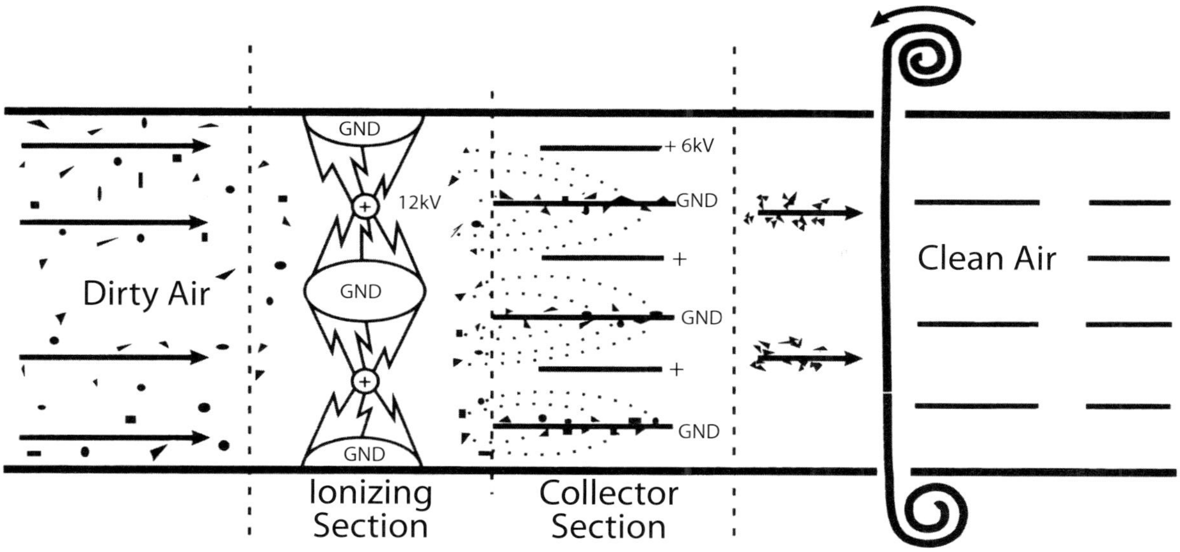

Figure 6.5. Principles of Agglomeration

stronger than that of the bond between the particles and the plate. These particles form agglomerates. As the agglomerates grow larger, their attraction to the plate is reduced and finally these agglomerates are re-entrained into the airstream. Agglomerates, since they are much larger than the small particles of which they are composed, are easily captured by lower-efficiency media air filters located downstream of the collector plates.

An agglomerator's collector cells are not intended to retain a heavy loading of the particles which they capture, and must always be used with afterfilters. The principles of agglomeration are shown in Figure 6.5.

Prefilters

The use of low-efficiency prefilters upstream of an electronic air cleaner is recommended for the following reasons:

1. The filters will act as air distribution panels to equalize the airflow through the electronic air cleaner.
2. They will capture large particles that could bridge the electrically charged plates causing a short circuit to deactivate the power supply. If shorting does occur, arcing caused by large particles can result in excessive power consumption and power pack failure.
3. By preventing arcing, the disturbing "snapping" noise made by the discharges is eliminated.
4. Elimination of arcing helps limit ozone generation.

Cleaning ESP Collector Cells

The removal of dust or mists that an electronic air cleaner has captured can be accomplished in three general ways:

1. With residential and light commercial units, contaminants can be washed off the collector plates. The typical unit allows the collector cell to be removed and washed using mild detergent and warm water.
2. For industrial models that are designed to collect coolant mist from cutting machinery, the mists collects on the plates, usually placed vertically in the unit, so gravity drips the collected liquid contaminant into a sump or directly back into the machinery. See Chapter 15: *"Industrial Contaminant Air Filtration Control Systems."*
3. In larger commercial and industrial applications, when contaminants agglomerate off the plates, they will be captured by a mechanical filter placed downstream of the collector section.

Washable Electronic Air Cleaners

Most all small residential and light commercial electronic air cleaners are the washable type. Their purpose is to capture and hold all the contaminants removed from the airstream. When the collector plates became loaded with dust, the performance of the electronic air cleaner is reduced (this is frequently indicated in the form of arcing between oppositely charged plates), the power to the electronic air cleaner is turned off and the plates are removed and washed.

Commercial Manual wash in place: Power to the electronic air cleaner is turned off, and an operator uses a power spray, and detergent solution, to clean the collector plates. The detergent and collected dirt drain to a floor pan sump connected to the sewer system. A final rinse with clear water may be used, after which the electronic air cleaner is allowed to drain and dry. The system is then put back in service.

Commercial Automatic wash in place: The electronic air cleaner is built with a properly located manifold of overhead and/or upstream spray nozzles so that when cleaning of the plates is required, the power is turned off and the cleaning system turned on.

The action of the cleaning liquid from the spray washes the collected dirt off the plates. The dirty liquid drains to the sewer system. When the plates have been dried, the electronic air cleaner can be put back in service (Figure 6.6).

Figure 6.6. Wash-in-Place Electronic Air Cleaner

Heavy Commercial/Industrial Mobile washers: In large installations, a separate washer, mounted on tracks in front of the electronic air cleaner, may be used to wash one section at a time. If this is done while the whole electronic air cleaner is off-line, the process is generally the same as for a built-in washer. If however, the electronic air cleaner must stay on-line at all times, each section may have its own separate manifold so that it can be isolated from the others. In this instance, each manifold incorporates a set of downstream dampers that are closed when cleaning is taking place. This prevents the by-pass of dirty air through a section that is being washed. Electronic air cleaners of this type are usually oversized so that they are operating at rated capacity when any one section is off-line being cleaned.

Afterfilters

Automatic roll filters perform exactly as they would in an airstream by themselves, except that the particles they are capturing are agglomerates and larger particles. As the roll media becomes loaded with dust, it is moved downward and clean media exposed at the top. The segment loaded with dirt is wound on the bottom spool which is changed when the roll of media is used. See Chapter 3, page 3.6.

Extended surface filters are used extensively as afterfilters for agglomerators. Since these extended surface filters have MERV 15 efficiency, their use can provide a fail-safe system. If the agglomerator fails to work for any reason, the afterfilter will still remove the contaminant in the airstream with high efficiency. One manufacturer also reports that the life of the afterfilter is considerably higher than it would be without the agglomerator. The agglomerates that blow off collector plates are much larger than unagglomerated contaminants. As a result, they do not block the passageways in the media of the extended surface filter and allow it to collect more contaminants before reaching the final pressure drop. Since a bank of extended surface filters is completely changed at one time, the flow across the filter bank is uniform, creating conditions for the agglomerator to operate at optimum efficiency.

Misapplication of Agglomerators

Under normal operating circumstances, cleaning of agglomerator collector cells may still be required as part of a yearly maintenance program. Liquid aerosols such as tars and oils will wet the surfaces of the collector plates, necessitating periodic cleaning.

Agglomerators should not be used where there are large concentrations of liquid aerosols in the air, such as in areas where there is heavy smoke since these may quickly load the plates and interfere with agglomerator operation. For such applications, washable collector assemblies are recommended.

Many self-contained units are tested and listed by the Association of Home Appliance Manufacturers (AHAM) ANSI/AHAM Standard AC-1 This standard is discussed in Chapter 7: *"HVAC Filter Testing."*

Service Requirements

Electronic air cleaners should be designed for easy servicing. Following is a list of some of the important design considerations:

1. Insulators should be made of high-quality arc-resistant material and should be mounted out of the airstream. If they are in the airstream, they will accumulate dust which may become a path for the leakage of the electrostatic charge.
2. The ionizer and collector cell sections should be designed for easy accessibility for effective cleaning and maintenance.
3. There should be some provision for isolating a section in large multiple-component units if this section becomes defective in any way.
4. A commonly used technique in industrial applications such as welding fumes is to utilize two cells back-to-back, which is called a double pass system.

Power Packs

Power packs are the heart of an electronic air cleaner. Power packs for heavy-duty industrial-commercial applications are designed to convert 110/220 volt single phase 50 to 60 Hz. power into 12,000/6000 volts (12 kV/6 kV) DC. One power pack may be adequate for a small installation, but two or more may be used on larger installations.

On small commercial and residential electronic air cleaners, including self-contained units, voltage may be only 8 kV for the ionizing section and 4 kV for the collecting section. Some units use 6 kV on both the ionizer section and collector section.

Applications Not Normally Suitable for Electronic Air Cleaners

Generally electronic air cleaners are not recommended for the following applications:

1. Conditions of very high humidity, especially air containing fine water droplets. The water droplets may cause arcing or even a shorting-out of the power pack. Industrial oil mist applications are acceptable for ESP's. See Chapter 15.
2. Systems where there can be a buildup of explosive gases.
3. Unattended equipment where periodic maintenance procedures are not assured.

Standards for Electronic Air Cleaners

UL 867 "Standard for Electrostatic Air Cleaners" covers electrical and fire safety requirements for electronic air cleaners used in the United States. Section 37 of the Standard addresses measurement of ozone emissions from air cleaners, but makes no comment about the air-cleaning ability of electronic air cleaners.

The Canadian Standard is designated CSA C22.2 No. 187-Mi986. It specifically states (Par. i.5) that the Standard does not specify the effectiveness of air cleaners with respect to the removal of airborne particles.

Other countries have their own standards.

6.8

Underwriters Laboratories Inc. Standard UL 867

In UL 867, the air cleaners are classified into five different categories:

- Duct type
- Fixed type
- Portable type
- Self-contained type
- Stationary type

For each of these types there are construction requirements relating to electrical safety, especially as it applies to the high voltage used in the ionizing and collecting sections of the electronic air cleaner.

No performance tests are required on any of the units, except the portable products for household use. UL Section 37 has been clarified for repeatability and reproducibility of ozone testing and states:

> "A portable air cleaning product for household use shall not produce a concentration of ozone exceeding 0.050 ppm by volume when tested as described in 37.2-37.7. A transitory concentration in excess of 0.50 ppm but less than 0.100 ppm is acceptable. However, the average of any five consecutive measurements taken 60 seconds apart shall be less than 0.050 ppm."[2] The test procedure, outlined in detail has many new criteria and NAFA encourages those desiring further test information to contact Underwriters Laboratories Inc. for this protocol."

Notes to Chapter 6

1. J.T. Handley, D.D. Smith, P.A. Lawless, D.S. Ensor and L.E. Sparks, "A Fundamental Evaluation of an Electronic Air Cleaner," presented at the 5th Annual International Conference on Indoor Air Quality and Climate, Toronto, Ont.
2. "Clarification for Ozone Testing of Electronic Air Cleaners and Ionizers, March 20, 2007." Underwriters Laboratories Inc., Research Triangle Park, Div. 12, Laboratory Drive, Research Triangle Park, NC 27709.

7

HVAC Filter Testing

Note: At the time of printing, there is a proposal through ISO Technical Committee 142 to have a recognized worldwide standard for testing HVAC air filters.

Filter Testing

Product evaluation or testing is an important consideration for any industry. In the case of air filters, the three major concerns are:

1. What will be the efficiency of the filter in removing the airborne contamination that is of interest to the user? Frequently there is interest in the ability of the filter to remove a specific size contaminant.
2. How much of this contaminant will it remove before maintenance is required?
3. What resistance (pressure drop) will the filter add to the system?

A separate concern is the safety of equipment in an HVAC system. In the United States, air filter fire safety is addressed in fire testing standards developed by Underwriters Laboratories (UL). Similar standards have been developed in other countries by various agencies.

Filter Performance Testing

When the air filter industry began, the proof of performance was found in the success of various installations. If a filter solved a ventilation problem, it did not matter how it was done. However, as the number of installations increased and as suppliers competed for the same business, it became important to have a standard method of testing so that there was a basis for evaluating filter products.

Even before the air filter industry developed its first filter test standard, the automotive industry found a test dust for evaluating carburetor intake filters. All the major automobile manufacturers had winter testing grounds, many of which were in Arizona. Thus the first test dust, Standardized Air Cleaner Test Dust, was originally derived from Arizona road dust.

ASHRAE Test Methods

There was a strong desire within the heating and ventilating industry to develop a standard for testing all types of air filters and to eliminate shortcomings in various test methods. The organization that finally developed the standard was ASHRAE. It had supported research into improvements in the earlier dust spot test methods, and the results of this research were reported at annual meetings of the Society and in related publications. Based on these findings and other information gathered from different sources, a committee was established to prepare a standard for testing air filters. The resulting standard was designated ANSI/ASHRAE Standard 52-68 and entitled Method of Testing Air Cleaning Devices Used in General Ventilation for Removing Particulate Matter. Its effective date was August 16, 1968.

ANSI/ASHRAE Standard 52-76 added two important changes to 52-68. These changes were: (1) To define the duct section holding the filter under test so that test filter pressure drops duplicated that of a bank of filters; and (2) To include International System of Units (SI) values along with the Inch-Pound (I-P) that had been previously used.

In the introduction to ANSI/ASHRAE Standard 52-76 the Committee responsible for developing the Standard noted:

> "The Project Committee in formulating the original Standard 52-68, updating it in November 1972, and ultimately revising this Standard visualized it as forming the test basis for product standards and for performance certification programs."[1]

A major disadvantage of ANSI/ASHRAE Standard 52-76 was the fact that the testing time required for determining ASHRAE atmospheric dust spot efficiency could be quite long. This seriously affected the cost of testing. Cooperative research done by different facilities indicated that a shorter testing method could be used and still provide the accuracy of ANSI/ASHRAE Standard 52-76. As a result, it was proposed that ANSI/ASHRAE Standard 52-76 be replaced by a new standard that allowed both the original and the shorter methods of testing. In 1992, ANSI/ASHRAE Standard 52.1-1992 was approved as such a standard. It allows testing by the procedure which was defined in 52-76, but also defines the shorter testing method which can used. In addition, it more precisely defines some procedures to get better reproducibility.

ANSI/ASHRAE Standard 52.1

As noted previously this Standard[2] has served the air filter industry since 1992. It is still referenced in many of the filter manufacturers' literature and in different related Standards and Regulations.

1. It requires both an atmospheric dust spot efficiency test and a synthetic dust weight arrestance test. However, it does not require an atmospheric dust spot efficiency test on filters that are acknowledged to have an average dust spot efficiency of less than 20%. Instead, it can be stated that this is the case.
2. It defines the atmospheric dust spot discoloration test method as "Efficiency" and the synthetic dust weight test as "Arrestance." Understanding how easily the word "efficiency" can be misinterpreted or misapplied, it requires the use of the descriptions "ASHRAE atmospheric dust spot efficiency" and "ASHRAE arrestance" to be used in all references to ANSI/ASHRAE Standard 52.1 test reports.
3. It reports on the amount of ASHRAE test dust captured by a filter until it reaches a designated final pressure drop and defines it as Dust Holding Capacity.

A brief description of the test protocol, test duct, related hardware and report form are described in detail in Appendix Five.

Note: At the time of printing, ASHRAE SSPC 52.1 Committee voted to incorporate 52.1 dust arrestance into 52.2 and retire 52.1.

Continuing ASHRAE Research

ANSI/ASHRAE Standard 52.1-1992 served a very useful purpose in providing an updated standard method for measuring the ability of a filter to capture a synthetic dust and to remove the discoloration fraction of fine dusts in the air. However, it did not give information on fractional efficiencies, namely the ability of a filter to remove particles of a certain size. Such information is very valuable in selecting a filter to provide a specific level of contamination control. Efforts by various laboratories to develop a new standard based on the efficiency of a filter on certain particle sizes gave inconclusive results.

Funds were requested from ASHRAE for a research project to be contracted to a professional research and development organization to determine if this type of testing was feasible. Research Triangle Institute was awarded a contract in April 1991.

In anticipation of completion of the work of Research Triangle Institute, ASHRAE approved the formation of Standards Project Committee (SPC) 52.2P Particle Size-Efficiency Procedures for Testing Air Cleaning Devices used in General Ventilation for Removing Particulate Matter. The purpose of this committee was to establish a test procedure for evaluating the performance of air cleaning devices as a function of particle size, to establish specifications for the equipment required to conduct the tests, to define methods of calculations from the data and to report the results obtained.

After much study by the committee and a period for public comment, ASHRAE approved a new standard 52.2-1999 Method for Testing General Ventilation Air-Cleaning Devices for Removal Efficiency by Particle Size.[3]

ANSI/ASHRAE Standard 52.2-1999

This new test is entirely different from ANSI/ASHRAE Standard 52.1.

1. It uses a laboratory-generated aerosol, potassium chloride (KCl), instead of atmospheric dust for the measurement of filter efficiency.
2. It does not measure discoloration efficiency but instead determines a filter's efficiency in removing 12 different size fractions of dust using a particle counter.
3. As in the case of ANSI/ASHRAE Standard 52.1 it uses ASHRAE test dust to load a filter between efficiency test runs.
4. It incorporates a variety of safeguards to guarantee the accuracy of the test results.
5. It requires multiple tests for every sample to assure statistical accuracy.
6. A set of particle size removal efficiency (PSE) curves at incremental dust loading is developed and together with an initial clean performance curve, is the basis of a composite curve representing the minimum performance in each size range.
7. Points on the composite curve are averaged and the averages are then used to determine the minimum efficiency reporting value (MERV) of the air cleaner. (See Figure 7.2).

The test duct and related hardware, test procedure, and test report form are described in Addendum 7.1 of this chapter.

EN 779:2002

The Comité Européen de Normalisation (CEN) developed (EN) 779. It is a standard developed from EN 779 and Eurovent 4/9:1997 to test general ventilation filters. The EN 779 test rig is similar to the ANSI/ASHRAE Standard 52.1 test rig and utilizes ASHRAE Test Dust to test for the course dust, and diethylhexylsebacate (DEHS) or its equivalent aerosolized with a Laskin nozzle which produces a particle size range of 0.2 - 3 µm to test for the fine dust. Filters with an efficiency higher than 98% on 0.4 µm particles are tested in accordance with EN 1822 (see Chapter 8: *"HEPA and ULPA Filter Testing"*).

Reported data for the test includes air flow, initial and final pressure drop, clean filter efficiency as a function of particle size, and then after the addition of ASHRAE Test Dust, average arrestance and dust holding capacity.

Filters are classified (on 0.4 µm particles) as G-type filters when efficiencies are lower than 40% and F-type filters when efficiency is higher than 40%.

Figure 7.1 shows the comparison of ANSI/ASHRAE Standard 52.1, ANSI/ASHRAE Standard 52.2 and EN 779 filter classifications.

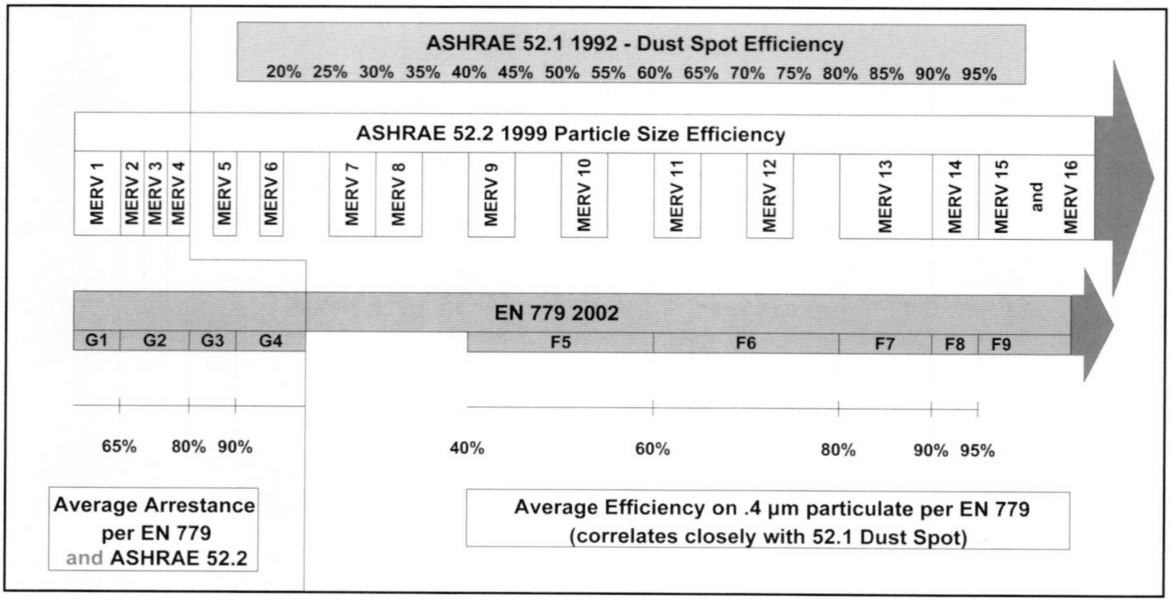

Courtesy Blue Heaven Technologies

Figure 7.1. Comparison of Filter Classifications; ANSI/ASHRAE 52.1 - 52.2 and EN 779

Third-Party Testing

Filter users often require that filter tests be performed by an independent third party testing laboratory. Most filter manufacturers have their products tested by one of these testing laboratories. In order to demonstrate credibility many end-users ask that the testing laboratory purchase the filter on the open market. This can be done if the filter is already in the commercial market but obviously cannot be done for first samples or prototypes. See Appendix Three for a list of independent laboratories capable of performing air filter testing.

Filter Certification

As presently conducted, third-party testing, while authoritative, provides data on one filter which may or may not be typical for all products of that type.

The particular merit of a certification program is that it provides a means of monitoring equipment performance and confirming that equipment performs as stated in the manufacturer's literature. Not only is a product tested initially, but it is subject to periodic testing to confirm that it is still in compliance.

NAFA Filter Certification Program

After more than two years of market research, studying the specific concerns of both supplier and user, NAFA has approved a product certification program described in the NAFA Product Certification Program Procedural Guide[5]. The highlights of the Guide are as follows:

1. The purpose of the certification program is to provide on-going certification performance claims made by air filter participants and to develop a data base of test results that provides performance data to the end user.

2. The program is voluntary with participation by air filter manufacturers.

3. Independent testing laboratories will be designated by NAFA as the facilities approved to perform NAFA Product Certification testing.

4. Filter samples will be taken from a warehouse stock of minimum quantity determined by the MERV number of the filter being tested.

Figure 7.2. MERV and Typical Applications (See also Figure 1.4)

ANSI/ASHRAE Standard 52.2 Minimum Efficiency Reporting Value (MERV)	Application Guidelines		
	Typical Controlled Contaminant	**Typical Applications and Limitations**	**Typical Air Filter/ Cleaner Type**
16 15 14 13	**0.3 - 1.0 μm Particle Size** All bacteria Most tobacco smoke Droplet nuclei (sneeze) Cooking oil Most smoke Copier toner	Hospital inpatient care General surgery Smoking lounges Superior commercial buildings	**Pocket Filters** 12 - 36 in. (300 to 900 mm) deep, 6 to 12 pockets. **Rigid Box/Cell Style Filter** 6 to 12 in (150 to 300 mm).
12 11 10 9	**1.0–3.0 μm Particle Size** Legionella Humidifier dust Lead dust Coal dust Auto emissions Nebulizer drops	Superior residential* Better commercial buildings Hospital laboratories	**Pocket Filters** 12 - 36 in. (300 to 900 mm) deep, 6 to 12 pockets. **Rigid Box/Cell Style Filter** 6 to 12 in (150 to 300 mm).
8 7 6 5	**3.0–10.0 μm Particle Size** Mold Spores Dusting aids Cement dust	Commercial buildings Better residential Industrial workplaces Paint booth inlet air	**Pleated Filters** Disposable, extended surface, 1 to 5 in (25 to 125 mm). **Cartridge Filters** Graded density cube or pocket filters, lofted synthetic media **Throwaway** Disposable synthetic media panel filters
4 3 2 1	**>10.0 μm Particle Size** Pollen Sanding dust Textile fibers Carpet fibers Paint overspray	Minimum filtration Residential Window air conditioners	**Throwaway Disposable** fiberglass or synthetic panel filters **Washable Aluminum** mesh, latex coated natural fibers, or foam rubber panel filters **Electrostatic** self charging (passive) woven polypropylene panel filter

* Residential system modification may be necessary for filters greater than MERV 8.

Source: ANSI/ASHRAE Standard 52.2–1999

5. Testing will be in accordance with ANSI/ASHRAE Standard 52.2 to final pressure drops of 1 in. w.g. (249 Pa) for filters with MERVs of 5 through 12 and 1.4 in. w.g. (349 Pa) for filters with MERVs of 13 through 16.
6. Filter testing will include removal efficiency by particle size method as detailed in ANSI/ASHRAE Standard 52.2 along with weight gain using the synthetic dust holding capacity.
7. Procedures are set up for the handling of filters not in compliance.
8. Filters in compliance will be issued a product certification symbol.
9. A procedure is defined for use when certified values are challenged.

This is not a complete listing of all the details and readers are referred to the NAFA Certified Product Procedural Guide for any additional information they may need.

Information Available from ASHRAE Filter Test Summary Sheets

A standard report form describing a filter test can provide a great deal of information about that filter to a buyer or seller. The information available from filter test sheets ANSI/ASHRAE Standard 52.2 is covered in Addendum 7.1 that follows this chapter. Testing organizations are not required to adopt the exact report form suggested by each Standard, but it is required that all the items shown in the suggested version be included in any version that they may use.

Filter Tests vs. Actual Filter Performance

Air filter testing is complex and no individual test adequately describes all filters. Ideally, performance testing of equipment should simulate operation under actual conditions and evaluate the characteristics important to the equipment user. Wide variations in the amount and type of particles in the air being cleaned make evaluation difficult. Another complication is the difficulty of closely relating measurable performance to the specific requirements of users. Recirculated air tends to have a larger proportion of lint than does outside air. However, performance tests should strive to simulate actual use as closely as possible ."[7]

The foreword to ANSI/ASHRAE Standard 52.2 contains the following notes:

<u>Electronic Air Cleaners</u>

Some air cleaners such as externally powered electrostatic precipitators (also known as electronic air cleaners) may not be compatible with the loading dust used in this test method. The dust contains very conductive carbon that may cause electric shorting, thus reducing or eliminating the effectiveness of these devices and negatively affecting their MERV. In actual applications the efficiency of these devices may reduce over time and their service life is dependent on the conductivity and the amount of dust collected.

<u>Passive Electrostatic Fibrous Media Air Filters</u>

Some fibrous media air filters have electrostatic charges that may be either natural or imposed upon media during manufacturing. Such filters may demonstrate high efficiency when clean and drop in efficiency during their actual use cycle. The initial conditioning steps of the dust-loading procedure described in the standard may affect the efficiency of the filter but not as much as would be observed in actual service. Therefore, the minimum efficiency during test may be higher than that achieved during actual use.

<u>Not an Application Standard</u>

Users should not misinterpret the intent of this standard. This is a test method standard, and its results are to be used to directly compare air cleaners on a standardized basis irrespective of their application. Results are also used to give the design engineer an easy-to-use basis for specifying an air cleaner. It is entirely possible that an industrial organization may use this test method as the basis for an application standard with, for example, different final resistances.

AHAM association of home appliance manufacturers

Room Air Cleaners

Room air cleaners may incorporate electronic air cleaners or they may use mechanical filters of various efficiencies to remove the objectionable airborne material. (See Chapter 6: *"Air Cleaners"*.) The Association of Home Appliance Manufacturers has developed a standard *Portable Household Electric Cord-Connected Room Air Cleaners* (ANSI/AHAM AC-1-2006).[8] The purpose of the standard is to establish "uniform, repeatable procedures and standard methods for measuring specific product characteristics of portable household electric cord-connected room air cleaners." It further states "The standard method provides a means to compare and evaluate different brands of portable household room air cleaners regarding characteristics significant to product use."

The test itself is based on a standard room (test chamber) whose characteristics are specified. Temperature is maintained at:

$$70 \pm 5°F\ (21 \pm 2.5°C)$$

and the relative humidity at:

$$40\% \pm 5\%.$$

Separate tests are run using cigarette smoke, air cleaner fine fraction test dust, and pollen. In all cases the room air cleaner is installed in the test chamber. The chamber is purged by recirculating test chamber air through a HEPA filter until the background count for airborne particulate sizes from 0.09 to 1.0 μm is less than 90 particles per cc as measured by an aerosol spectrometer. Then one of the three contaminants is introduced into the room and distributed uniformly by a circulating fan operating for 1 minute.

During this cycle the room air cleaner is not operating. Dust particle concentration readings within the room are taken at every 2 minute intervals until nine data points have been taken. During this period natural decay (reduction of concentration of contaminant) has occurred because of gravity or adherence to the surface of the walls or content of the room. This data can be converted into a natural decay curve.

The room is then purged and again the same test contaminant is introduced and dispersed throughout the room. However, this time the room air cleaner is operating. Again a series of counts are taken at 2 minute intervals and converted into a decay curve. Depending on the air cleaning efficiency and quantity of air being handled by the room air purifier, the second test should show a more rapid reduction in particle concentration within the room.

The performance of the room air cleaner is given in terms of a Clean Air Delivery Rate (CADR). Clean air delivery rate is defined by the equation:

$$CADR = V\,(k_e - k_n)$$

Where:

V = volume of test chamber (ft^3)

k_e = measured decay rate

k_n = natural decay rate

If the measured decay rate is high compared to the natural decay rate, the CADR will be correspondingly higher. In turn, the higher the CADR, the greater the ability of the room air cleaner to reduce the concentration of contaminants.

Fire Resistance

The protection of buildings and contents against fires has resulted in the creation of organizations dedicated to this specific problem. One such organization is Underwriters Laboratories Inc., self described as "an independent, not-for-profit organization testing for public safety". It publishes a Buildings Material Directory.[9] In this directory it further describes its mission:

Underwriters Laboratories Inc. (UL) provides a service for the classification or evaluation of products with respect to (a) the property of the product, (b) a limited spectrum of hazards to life or property, (c) suitability of the product for certain uses, or (d) such other conditions as UL may consider desirable. Such service is referred to as the "Classification Service" and products covered under the Service are referred to as being "Classified" when bearing a "Classification Marking." Air Filters are listed under "AIR CONDITIONING SYSTEMS (AIXZ)."

Filter Units. Air (AJZV)
These are filters used for the removal of dust and other airborne particles from air circulated mechanically in equipment and systems installed in accordance with the Standard of the National Fire Protection Association for Installation of Air Conditioning and Ventilating Systems of other than Residence Type NFPA No. 90A and for Installation of Residence Type Warm Air and Air Conditioning Systems NFPA No. 90B for air conditioning, cooling heating, and/or ventilating.

The basic standard used to investigate products in this category is UL 900 "Test Performance of Air Filter Units."

UL Standard 900

As a result of acceptable testing under UL 900[10] a filter can be listed in one of two categories:

Class 1 Units—those that, when clean, do not contribute fuel when attacked by flame and emit only negligible amounts of smoke.

A Class 1 air filter unit shall not produce flame or sparks when subjected to the Flame-Exposure Test.

The Class 1 air filter unit shall not cause the development of an area of more than 2.5^2 in. (16.1 cm^2) as measured below the smoke-density time curve.

Class 2 Units—Those that, when clean, burn moderately when attacked by flame or emit moderate amounts of smoke, or both.

A Class 2 air filter unit shall not produce flame or extensive (25 or more) sparks which are sustained beyond the discharge end of the test duct when subjected to the flame-exposure test and shall not cause the development of an area of more than 9 ^{2} in. (58 cm^2) as measured below the smoke-density time curve.

All filters to be tested under UL 900 are subjected to the Flame-Exposure Test. Class 1 filters with face areas larger than 20 x 20 inches are subjected to the Spot Flame test of the unit and of components not in the airstream during the flame-exposure test.

If a filter normally is adhesive coated, prior to the test, this filter must be totally immersed in the correctly formulated adhesive material and allowed to drain in the vertical position for 24 hours before being tested.

Flame-Exposure Test

In the Flame-Exposure Test, the filter is placed in a precisely defined duct section attached to a blower capable of delivering a specific amount of air against the resistance offered by the filter under test and by the duct system. This test air duct velocity is standardized at 220 fpm (1.12 m/s) regardless of the flow rate for which the original filter was designed. A multiple-nozzle gas burner is located 18 in. (460 mm) upstream of the test filter. The gas-burning rate is 4000 Btu/min (70.3 KW) ±5%. A duct section approximately 8 ft (2.44 m) long is downstream of the filter. The outlet end of this section contains two opposite holes through which a beam of light located above the duct can be focused on a photoelectric cell below the duct. This is the device used to measure the amount of smoke generated, by recording the reduction in the intensity of the beam of light as it passes through any smoke generated during the test.

Spot-Flame Test

The Spot-Flame Test apparatus consists of a precisely described Bunsen burner adjusted to produce a clearly defined oxidizing flame 6 in. (152 mm) high. A holding device supports the samples of each component part or filtering medium or an air filter 3 in. (76 mm) above the top of the burner barrel where it is exposed for 1 minute. Observations are made and recorded of any flaming of the test sample or production of sparks.

The technique for measuring the amount of smoke generated during the test is outlined in detail. The upstream face of a Class 1 filter shall not continue to flame after removal of the test flame.

Clean vs. Dirty Filters

The significance of UL filter ratings has always been questioned because the tests were performed on clean filters. Critics have pointed out that much of the dust captured by filters in air conditioning systems includes lint and other materials that will burn or at least will create smoke when exposed to high temperatures or flame. Since this is equally true for either a Class 1 or a Class 2 filter, the question has been raised as to the value of a Class 1 over a Class 2 filter.

In May 1983 Underwriters Laboratories revised paragraph 1.2 of UL 900 to read:

> "Since the combustibility and smoke generation of an air filter unit, after a period of service, will depend upon the nature and quantity of the material collected by the filter, the test requirements of this standard, for classification purposes, apply only to air filter units in a clean condition. *Consequently, when filters are susceptible to the accumulation of combustible deposits. It is intended that maintenance and inspection processes should be followed as proposed in Appendix B of NFPA 90A."*

Note: We have italicized the material which was added.

Appendix B of NFPA 90A states the following regarding filter maintenance:

B-3 Filters

B-3.1 All air filters should be kept free of excess dust and combustible material. Unit filters should be renewed or cleaned when the resistance to airflow has increased to two times the original resistance or when the resistance has reached a value of recommended replacement by the manufacturer. Provide a suitable draft gauge for the purpose...

B-3.2 When renewing filters, care must be taken to use proper type and size to avoid gaps between filter sections, mounting frames, or hardware. Damaged filter sections or media should not be used.

B-3.3 Filters designed and manufactured to be thrown away after use should never be cleaned and reused.

B-3.4 Care should be exercised in the use of liquid adhesives. Use of an adhesive of low flash point would create a serious hazard.

B-3.5 Electrical equipment of automatic filters should be inspected semi-annually, observing the operation cycle to see that the motor, relays, and other controls function as intended. Drive motors and gear reductions should also be inspected at least semi-annually and lubricated when necessary.[11]

Notes to Chapter 7

1. ANSI/ASHRAE Standard 52-1976 Gravimetric and Dust Spot Procedures for Testing Air Cleaning Devices used in General Ventilation for Removing Particulate Matter American Society of Heating, Refrigerating and Air-Conditioning Engineers Inc., Atlanta, GA 30329.
2. ANSI/ASHRAE Standard 52.1-1992 Gravimetric and Dust Spot Procedures for Testing Air Cleaning Devices used in General Ventilation for Removing Particulate Matter American Society of Heating, Refrigerating and Air-Conditioning Engineers Inc., Atlanta, GA 30329.
3. ASHRAE Issues a Call for Members Air Media Aug./Sept. 1992, p. 5.
4. ANSI/ASHRAE Standard 52.2-1999 Method for Testing General Ventilation Air-Cleaning Devices for Removal Efficiency by Particle Size, American Society of Heating, Refrigerating and Air Conditioning Engineers Inc., 1791 Tullie Circle NE, Atlanta GA 30329.
5. NAFA Product Certification Program Procedural Guide, National Air Filtration Association, 291 Independence Blvd., Pembroke Four Suite 515, Virginia Beach, VA 23462.
6. Ibid. Appendix E—Cross—references and Application Guidelines.
7. 2004 ASHRAE Handbook, Systems and Equipment, Chapter 25 Air Cleaners for Particulate Contaminants p. 24.2.
8. Portable Household Electric Cord-Connected Room Air Cleaners, Standard ANSI/AHAM AC1-2006, Association of Home Appliance Manufacturers, 1111 19th St. NW (Suite 402), Washington DC 20036.
9. Building Materials Directory, Underwriters Laboratories Inc., 222 Pfingsten Rd., Northbrook IL 60062.
10. ANSI/UL 900-1995 Standard for Air Filter Units. Underwriters Laboratories Inc., 333 Pfingsten Rd., Northbrook IL 60062.
11. Appendix B of NFPA 90A

Addendum 7.1

A Brief Description of the ANSI/ASHRAE Standard 52.2 Test Method

Note: ANSI/ASHRAE Standard 52.2 is under "continuous maintenance" by ASHRAE Standing Standard Project Committee (SSPC) 52.2, and may be revised from this description.

The ANSI/ASHRAE Standard 52.2 Test Method[1] determines the ability of an air cleaner (air filter) to remove particles of a specific size from an air stream. Unlike previous dust spot discoloration efficiency tests that use atmospheric dust, 52.2 deliberately excludes all atmospheric dust by filtering test air through a HEPA filter before loading this test air with a synthetic test aerosol, potassium chloride. All testing is performed with the test duct under positive pressure.

Its only association with ANSI/ASHRAE Standard 52.1 is that it uses ASHRAE Synthetic Dust to incrementally load filters under test to the final pressure drop recommended by the manufacturer. Although there is a synthetic dust loading, there is no measure of ASHRAE arrestance in ASHRAE 52.2. Despite this statement, ASHRAE 52.1 arrestance is one of the pieces of data required before "coarse air filters" can be assigned MERVs of 1 to 4. (MERVs are discussed in more detail later in this addendum).

The two specific items covered by ASHRAE 52.2 are the ability of the air cleaner to remove particles of a given size from an air stream and the pressure drop of the clean filter for different air flows through it.

Filters must be tested at an air flow rate not less than 0.22 m/s (472 cfm) and no more than 1.4 m/s (3000 cfm). If not stated otherwise, filters will be tested at an air flow rate of 2.50 m/s (492 fpm) Likewise, if not stated otherwise, filters will be tested to a final pressure drop of 350 Pa (1.4 in w.g.).

General Description

There is strong emphasis on the standardization and calibration of components. The apparatus must undergo 13 qualification tests before filter testing begins. Theoretically a filter tested by ASHRAE 52.2 in one laboratory should have the same performances values when tested in any other correctly-operating laboratory.

The test setup uses an optical particle counter using incandescent light (or a laser source) to count the different size particles in the air upstream and downstream of the filter under test. The result is divided into 12 different ranges in μm. These size ranges are shown below:

Range	Lower Limit μm	Upper Limit μm	GMPS*
1	0.30	0.40	0.35
2	0.40	0.55	0.47
3	0.55	0.70	0.62
4	0.70	1.00	0.84
5	1.00	1.30	1.14
6	1.30	1.60	1.44
7	1.60	2.20	1.88
8	2.20	3.00	2.57
9	3.00	4.00	3.46
10	4.00	5.50	4.69
11	5.50	7.00	6.20
12	7.00	10.00	8.37

**GMPS = Geometric Mean Particle Size*

Addendum 7.1.2

Figure 7.1.1. Components of an ANSI/ASHRAE Standard 52.2 Test Duct

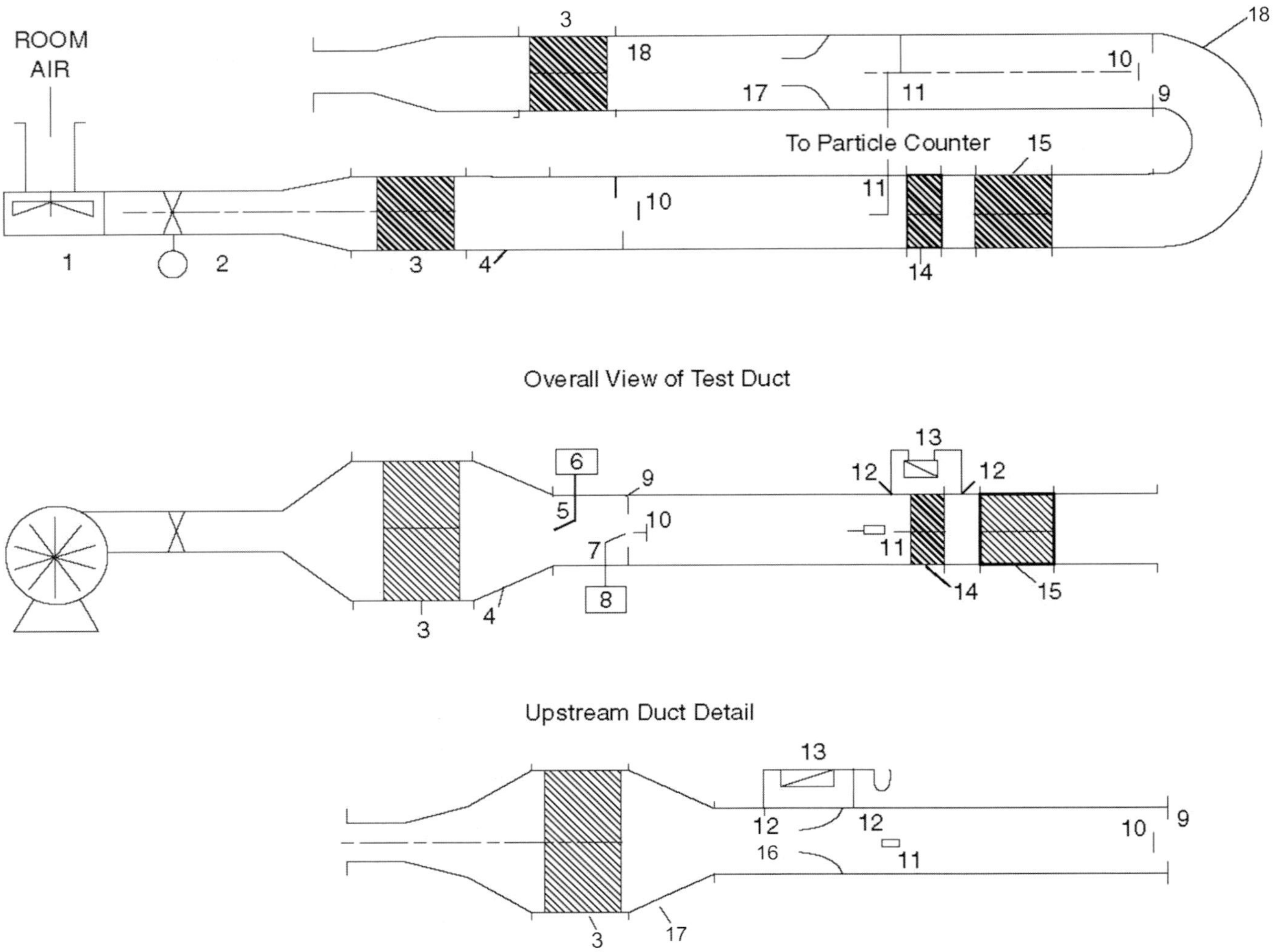

1. Blower
2. Flow Control Valve
3. HEPA Filter Bank
4. Transition, if required
5. Aerosol Injection Tube
6. Aerosol Generator
7. Dust Feed Pipe
8. Dust Feeder
9. Mixing Orifice
10. Perforated Diffusion Plate
11. Location of Sample Probe
12. Static Tap
13. Manometer
14. Air Cleaner Device and Transition, if any
15. Final Filter (installed only during dust loading
16. Main Flow Measurement Nozzle
17. Transition, if needed
18. Bend, optional

Test Equipment

Components of the ASHRAE test duct are shown in Figure 7.1.1. More complete details of all the components will be found in the ASHRAE Standard 52.2 itself.

Test Procedure

The test procedure (after the system has met all 13 qualification measurement requirements) consists of the following steps:

1. Determine the pressure drop of the filter under test at 50%, 75%, 100% and 125% of the test airflow rate.
2. Determine particle size efficiency (PSE) of the filter at the specified airflow test rate. At least 3 runs are made for each PSE curve. More may be required to achieve statistical accuracy.
3. The particle counts upstream and downstream of the filter are recorded for each test point of each size group. Filter efficiency at each point is calculated using the formula:

$$\text{Eff.} = (1 - Cd/ Cu) \times 100$$

Where:

Cu = Upstream particle count
Cd = Downstream particle count

4. The testing procedure is as follows:

   a. Determine the PSE of the clean filter
   b. "Condition" the filter by loading it with 30 grams of synthetic test dust or until the pressure drop increase is 10 Pa (0.04 in. w.g.), whichever comes first.
   c. Determine the PSE of the "conditioned filter"
   d. Load the filter with synthetic dust until the pressure drop has increased by 1/4 of the difference between the pressure drop of the clean filter and the final pressure drop to which the test is run.
   e. Determine the PSE of the filter.
   f. Repeat the loading with synthetic dust until the pressure drop reaches 1/2 of the difference.
   g. Determine the PSE of the filter.
   h. Repeat steps "f" and "g" at 3/4 of the resistance difference and at the final specified resistance
   i. As a result of these tests there will be six sets of PSE measurements
   j. Develop an efficiency curve for each set of measurements on special paper that breaks the 12 size ranges into three segments of four data points each
   k. Develop a composite minimum efficiency curve by plotting the minimum PSE in each of the 12 size ranges shown on the plots of each curve
   l. Average the four data points for this minimum PSE in each size range group and report these averages as E1, E2 and E3. Many will find that this is the information they need and stop the examination of results at this point
   m. A final test result simplification step can be taken by using the averages mentioned previously to generate one number called the minimum efficiency reporting value (MERV). The MERV is arbitrarily defined but can be selected by reference to Figure 7.1.2 *Minimum Efficiency Reporting Value (MERV) Parameters*

Figure 7.1.2. Minimum Efficiency Reporting Value (MERV) Parameters

Standard 52.2 Minimum Efficiency Reporting Value (MERV)	Composite Average Particle Size Efficiency, % in Size Range, µm Range 1 0.30 - 1.0	Range 2 1.0 - 3.0	Range 3 3.0 - 10.0	Average Arrestance, %
1	n/a	n/a	$E_3 < 20$	$A_{avg} < 65$
2	n/a	n/a	$E_3 < 20$	$65 \le A_{avg} < 70$
3	n/a	n/a	$E_3 < 20$	$70 \le A_{avg} < 75$
4	n/a	n/a	$E_3 < 20$	$75 \le A_{avg}$
5	n/a	n/a	$20 \le E_3 < 35$	n/a
6	n/a	n/a	$35 \le E_3 < 50$	n/a
7	n/a	n/a	$50 \le E_3 < 70$	n/a
8	n/a	n/a	$70 \le E_3$	n/a
9	n/a	$E_2 < 50$	$85 \le E_3$	n/a
10	n/a	$50 \le E_2 < 65$	$85 \le E_3$	n/a
11	n/a	$65 \le E_2 < 80$	$85 \le E_3$	n/a
12	n/a	$80 \le E_2$	$90 \le E_3$	n/a
13	$E_1 < 75$	$90 \le E_2$	$90 \le E_3$	n/a
14	$75 \le E_1 < 85$	$90 \le E_2$	$90 \le E_3$	n/a
15	$85 \le E_1 < 95$	$90 \le E_2$	$90 \le E_3$	n/a
16	$95 \le E_1$	$95 \le E_2$	$95 \le E_3$	n/a

Filter Test Report Form

ASHRAE 52.2 does not require any specific report format, but it makes the following statement in section 11.2.

"The summary section of the performance report shall include the following information:

1. Name and location of the test laboratory
2. Date of the test
3. Test operator's name(s)
4. Brand and model number of the particle counting and sizing device(s)
5. Air cleaner manufacturer's name (or name of the marketing organization, if different from the manufacturer)
6. How the sample was obtained
7. Description of the test air cleaner, including:

a. Brand and model number

b. Physical description of construction (e.g. extended surface - number of pockets or number of pleats; pleated panel number and depth of pleats)

c. Face dimensions and depth

d. For fiber media air cleaners:

i. Type and color of media

ii. Effective media area

iii. Type and amount of dust adhesive, if known

iv. Electrostatic charge, if known

a. Any other pertinent descriptive attributes

8. Operating data as stated by the manufacturer

   a. Test conditions for reporting purposes: airflow rate and final resistance
   b. Initial and final resistances
   c. Any other operating data furnished

9. Test data

   a. Test air temperature and relative humidity
   b. Airflow rate
   c. Loading dust weight*
   d. Type of test aerosol

* Note: This can easily be confused or misunderstood as "Dust Holding Capacity." This is not DHC, rather the total amount of dust fed to final resistance.

DHL data is not reported on as part of 52.2 test method.

10. Results of resistance testing

    a. Initial resistance
    b. Final resistance

11. Performance curves

    a. A curve in Figure 11-1b format of air cleaner resistance when clean vs. airflow rates from 50% to 125% of test flow.
    b. A curve in Figure 11-1 c format of PSE for the clean device and for the device at each of the five loading stages.
    c. A minimum PSE composite curve in Figure 11-ic format whose data points are the lowest PSEs from the six measurements in each particle size range from the curves of test results (K.2 above)

12. Minimum Efficiency Reporting Value (MERV)

    a. The average of the minimum PSE of the four size ranges from 0.30 mm to 1.0 mm (E_1)
    b. The average of the minimum PSE of the four size ranges from 1.0 µm to 3.0 µm (E_3)
    c. The average of the minimum pSE of the four size ranges from 3.0 µm to 10.0 µm (E_3)
    d. MERV for the device.

Note to Addendum 7.1

1. ASHRAE Standard ANSI/ASHRAE Standard 52.2-1999 Method of Testing General Ventilation Air Cleaning Devices for Removal Efficiency by Particle Size, American Society of Heating, Refrigerating and Air-Conditioning Engineers Inc., 1791 Tullie Circle NE, Atlanta GA 30329.

8

HEPA and ULPA Filter Testing

Introduction

Critical air cleaning applications require the very high efficiencies that HEPA and ULPA filters are capable of achieving. So critical are most of the applications for these filters, that their performance has to be assured. This can only be done by individually testing each filter before its use. Such a test has to be nondestructive so that the filter performance is not impaired in any way.

The criteria required for a device to test HEPA filters are:

1. It should generate an aerosol whose size is determined to be the most penetrating particle size (MPPS) for HEPA filter media, that is, the aerosol size should be one for which HEPA filters have the lowest efficiency. This was originally calculated to be 0.3 micrometers but now is filter and flow rate dependent.

2. This aerosol should be monodispersed. That is, all particles should be the same size.

3. There should be a way of confirming the particle size.

4. There should be a means of accurately measuring the concentration of the test aerosol upstream and downstream of the filter under test.

This determines penetration which is then converted to efficiency.

All of these characteristics were thought to be found in the DOP efficiency tester when it was originally developed.

The Hot DOP Efficiency Test Q107 Penetrometer (MIL-STD-282)

Many changes and improvements have been made since basic information generated in the early 1940s allowed the development of the first dioctylphthalate (DOP) Efficiency Tester. DOP is an oily liquid. Its correct chemical name is diethylhexylphthalate (DEHP). The DOP efficiency test is sometimes referred to as the "Hot DOP test" to distinguish it from and in-place test in which a cold challenge aerosol is used. Originally, the DOP efficiency tester was sized for testing filter media and gas mask filters at only a few cfm (L/s). The time and place of the first successful upscaling from a few cfm to 1000 cfm is not known. The best guess is that it was done in the late 1940s at Edgewood Arsenal, where large collective protectors were developed using the earliest version of the HEPA filter. Collective protectors are filters for defense shelter ventilation systems. They remove chemical, biological and radiological (CBR) warfare agents.[1]

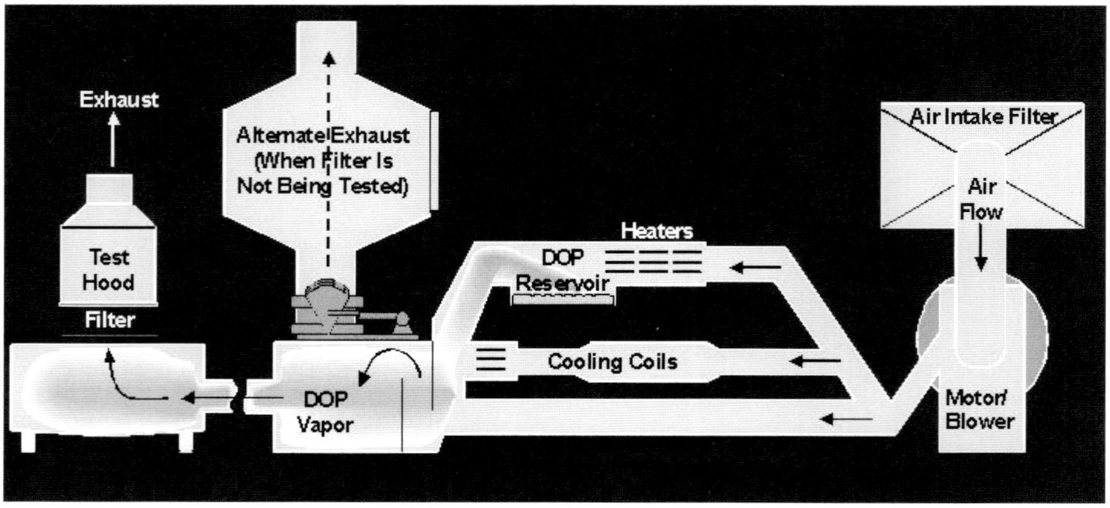

Figure 8.1. Schematic Diagram of DOP Tester

Military Standard (MIL-STD) 282 describes the method to test HEPA filters known as the "hot DOP" test. A schematic drawing of a typical DOP tester is shown in Figure 8.1. Air entering is filtered through HEPA filters to remove airborne dust. The blower discharges air into three channels. In the first, air may be heated and is then passed over hot liquid DOP where it picks up DOP vapor (gas). The second channel is air at room temperature that mixes with and cools the DOP vapor laden air and causes the DOP vapor to condense into a liquid aerosol, ideally 0.3 µm in diameter. The third channel contains dilution air. This dilutes the mixture containing the DOP aerosol, so that the concentration is reduced below that where agglomeration is likely to occur.

The flow from all three channels goes into a mixing and aging chamber where the formation of the DOP aerosol is stabilized. During a test, the DOP is passed into a chamber in which the filter under test is clamped. At the end of the test, the filter is removed, the test air vented back to the atmosphere, and the test chamber purged. Before the start of testing, a sample of air from the aging chamber is drawn into the "owl," an optical device that measures the particle size of the aerosol by the angular degree of light polarization.

A DOP tester measures penetration, that is, the percent of DOP that passes through the filter. Efficiency is 100% minus the penetration. Air is sampled from the plenum on the downstream side of the filter under test and piped to the photometer where penetration is measured.

Before testing is begun, separate HEPA-filtered air is drawn through the photometer so that the particle reading is zero. With the setting at minimum sensitivity, DOP laden test air is drawn into the photometer and the device adjusted to show 100% penetration. Then the penetrometer and connected sampling lines are purged, the device set to maximum sensitivity and filter tests are run.

When a HEPA filter is under test, it is encapsulated, i.e., held in place in such a way that all DOP penetration channels are measured. This includes not only the media, but also the frame and any gaskets that may be used. The direction of airflow during the test determines which portion of the frames and gaskets is challenged.

During testing, the pressure drop of the HEPA filter at rated capacity is also measured. Each filter is individually labeled and the penetration, pressure drop [in. w.g., (Pa)] and test air flow [cfm (m^3/s)] recorded on the filter.

Concern about the fact that DOP has been identified as a potential carcinogen has led to use of alternative materials. In some test facilities DOP has been replaced by materials considered less hazardous, such as polyalfaolefin (PAO).

Leakage vs. Penetration (Two-Flow Testing)

During the very early stages of development of gas mask filters it was recognized that penetration was due not only to the basic nature of the filter paper but also to "pinholes" which sometimes occurred in the media. These pinholes could be microscopic holes or they could be thin spots in the filter media. It was also determined that while penetration through media was essentially constant through the normal range of test velocities, the relative penetration due to pinholes increased as the test velocity was reduced. As a result, testing of media at two different velocities showed whether the penetration was due to the nature of the media or to pinholes created during media manufacture. If the penetration was essentially less or the same at the lower velocity, there were no pinholes. If it increased as the velocity was reduced, the penetration due to pinholes became more significant.

When HEPA filters came to be used in laminar flow devices, it was important that there be no leaks (see Chapter 5: *"HEPA, ULPA and Super ULPA Filters"*). One way of testing for such leaks was to run a DOP test at 100% and 20% of rated capacity. If the penetration was essentially the same in both tests there were no leaks. If it was greater when tested at 20% of rating, there was leakage either through pinholes in the media or in the seal of the filter pack to the frame. Some specifications define the maximum difference allowed between the two readings.

In-place Leak Testing; Cold DOP Test

Testing of a filter at 100% and 20% of rating can be used to determine the presence of leaks but it does not identify how many or where they are located. To do this requires a method of probing the filter face.

In testing for a leak, one is no longer concerned about most penetrating particle sizes. Leaks occur through openings that are very much larger than this size. For this reason it is not necessary to use the large and complicated apparatus required for the generation of uniform liquid aerosol particles. Instead, compressed air blown through cold DOP or all other suitable challenge liquid using specially designed Laskin nozzles creates an aerosol that is satisfactory. This test is sometimes referred to as the "cold DOP test," even if DOP is not the challenge used. A comparison of hot and cold DOP particle size is shown on Figure 8.2.

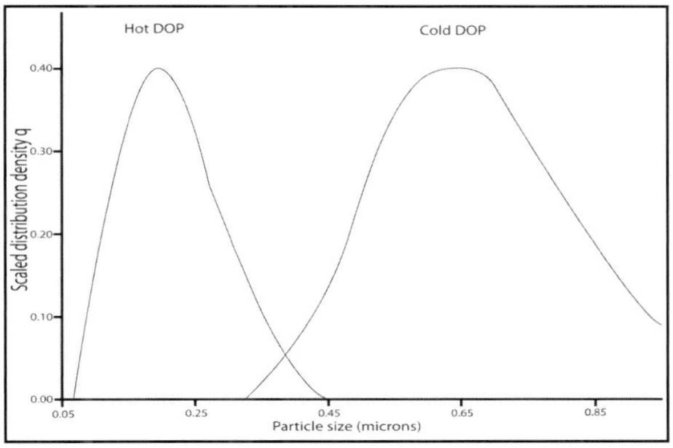

Reference: EN1822:1998

Figure 8.2. Particle Size Distribution of Hot & Cold DOP

The IEST Recommended Practice *CC034.*[2] *HEPA and ULPA Filters Leak Tests* includes leak testing of HEPA filters. This is intended primarily for cleanroom applications.

Where necessary, leak re-testing can be done at a job site just prior to the installation of the filters. Some contract specifications require this. By testing just before installation, one is assured that the filter has not been damaged between the time of factory testing and arrival at the site. For many applications, leak testing is required after installation (in-situ) to ensure the integrity of the filters and the installation itself. In critical applications such as biological safety cabinets, periodic leak testing is also required.

Particle Counters

A particle counter is a sophisticated monitoring instrument that counts the amount of particulate contamination. Modern particle counters detect contaminants from the air, a surface or a liquid. Some examples of these particles can be anything from oil, metal shavings, dust, smoke, ash, mold, or other biological contaminants. This information is very useful to professionals for determining how pure the air is for breathing, water treatment, producing food & drugs or microchips in a cleanroom.

Particle counters proved invaluable for monitoring the concentration of dust in cleanrooms during the Apollo space program. The earliest design used incandescent light bulbs as their light source. Despite improvements in light detectors and in internal optics, the smallest size particle which could be detected accurately by these devices had a diameter of 0.3 µm. Efforts to detect particles smaller than this usually resulted in electronic noise that made readings unreliable. The particle counter was improved when lasers were used in place of incandescent light. (See Figure 8.3.) The high-intensity light beam created by the laser, along with the coherent nature of the laser light itself, allowed for the detection and counting of smaller airborne particles.

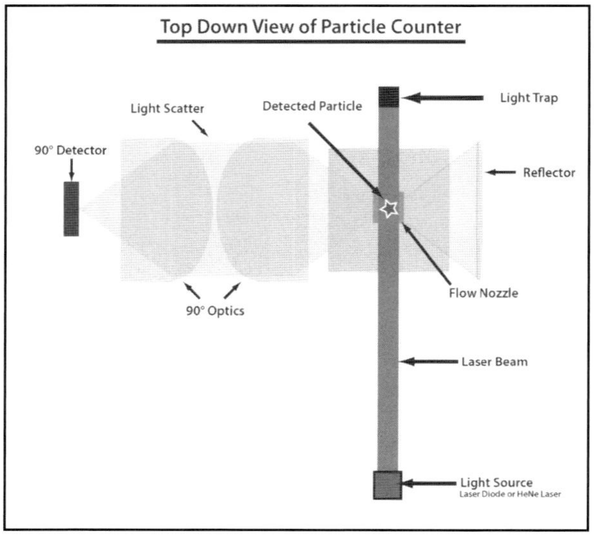

Photo Courtesy Wikipedia

Figure 8.3. Diagram Particle Counter

A particle counter can be as simple as a handheld air monitoring instrument or be as advanced as a remote or multiplex manifold sequencing clean room systems used for entire facility monitoring. Depending on the model and type, laser particle counters can detect as small as 0.1 µm.

Condensation Nucleus Particle Counters

In a laminar-flow, alcohol-based Condensation Nucleus Particle Counter (CNC), an aerosol sample is drawn continuously through a heated saturator in which alcohol vaporizes and is diffused into the sample stream. Together, the aerosol sample and alcohol vapor pass into a cooled condenser where the alcohol vapor becomes supersaturated and ready to condense. Particles present in the sample stream serve as condensation nuclei. Once condensation begins, particles that are larger than a threshold diameter grow quickly into larger droplets and pass through an optical detector where they are counted. CNC can detect particles as small 2.5 nanometers (0.0025 micrometers).

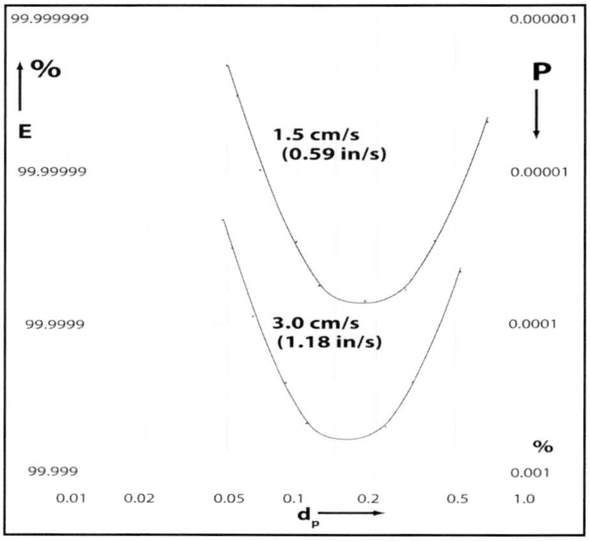

Reference: EN 1822:1998.

Figure 8.4. Particle Size Efficiency (E) and Penetration (P) of an ULPA filter medium as a function of particle diameter DOP at two different media velocities.

Higher Efficiency Air Filters

Filters with efficiencies higher than the HEPA filter have been available since the 1970s. Based on the DOP test method, these filters had efficiencies of 99.999%. The advent of microelectronics and its associated need for cleaner environments in which to manufacture microchips created a demand for these filters with even higher efficiencies. Today they are known as ULPA (Ultra Low Penetration Air) filters. Newer research tools, especially the laser particle counter and condensate particle counters along with new challenge aerosols such as poly styrene latex spheres (PSL) have been used to evaluate HEPA and ULPA filter performance. Laser particle counters were used to study the DOP filter tester. They revealed that:

1. The DOP aerosol generated in a DOP tester is not uniformly 0.3 µm in diameter. (Not homogenous.)

2. The most penetrating particle size (MPPS) was not 0.3 µm.

3. The most penetrating particle size for any media varied with the media velocity. As the media velocity decreased, the most penetrating particle size also decreased (see Figure 8.4).

The DOP tester does not have the sensitivity to measure the efficiency of an ULPA filter with any accuracy. To compensate for this fact, IEST has developed a Recommended Practice[3] 1007.1 for testing ULPA filters. The test procedure covers the testing of filters for particle penetration and pressure drop. Using this procedure it is possible to measure penetration ranges of from 0.001% to 0.00001% using test aerosols in the range of 0.1 to 0.2 µm.

Filters with efficiencies higher than ULPA filters are available but their use is limited. Known generically as Super ULPA (Super Ultra Low Penetration Air) filters, such filters are usually 99.9999% efficient on 0.12 µm particles. Buyer-seller agreements often specify a protocol that includes testing the media to be used and leak testing the completed filters. Figure 8.5 shows IEST categories of HEPA and ULPA filters used and test methods for each category.

EN 1822:1998

EN 1822:1998 is the standard used for HEPA and ULPA filter testing in Europe. The test defines characteristics for:

1. classification
2. performance testing
3. leak-finding
4. collection efficiency determination

Figure 8.5. IEST-RP-CC--1.4 - Recommended test and minimum rating for filter Types A through K

Filter Type	Penetration Test		Scan Test (see note)		Comments	Minimum Efficiency Rating
	Method	Aerosol	Method	Aerosol		
HEPA (Type A)	MIL-STD-282	Thermal DOP	None	None		99.97%
HEPA (Type B)	MIL-STD-282	Thermal DOP	None	None	Two-flow leak test	99.97%
HEPA (Type C)	MIL-STD-282	Thermal DOP	Photometer	Polydisperse DOP/PAO		99.99%
HEPA (Type D)	MIL-STD-282	Thermal DOP	Photometer	Polydisperse DOP/PAO		99.999% at 0.3 µm
HEPA (Type E)	MIL-STD-282	Thermal DOP	None	None	Two-flow leak test	99.97%
ULPA (Type F)	IEST-RP-CC007	Open	Particle Counter Photometer**	Open		99.999% at 0.1-0.2 or 0.2-0.3 µm
Super ULPA (Type G)	IEST-RP-CC0021*	Open	Particle Counter	Open		99.9999% at 0.1-0.2 or 0.2-0.3 µm
HEPA (Type H)	IEST-RP-CC007	Open	Photometer	Polydisperse DOP/PAO		99.97% at 0.1-0.2 or 0.2-0.3 µm
HEPA (Type I)	IEST-RP-CC007	Open	None	Open	Two-flow leak test	99.97% at 0.1-0.2 or 0.2-0.3 µm
HEPA (Type J)	IEST-RP-CC007	Open	Particle Counter or Photometer	Polydisperse DOP/PAO		99.99% at 0.1-0.2 or 0.2-0.3 µm
ULPA (Type K)	IEST-RP-CC007	Open	Particle Counter or Photometer	Polydisperse DOP/PAO		99.995% at 0.1-0.2 or 0.2-0.3 µm

**Filter medium tested at most-penetrating particle size (MPPS) prior to filter assembly. These filters are not tested for overall penetration.*
***This filter (Type F) may be leak tested with either photometers or particle counters. When hand scanning in the field extra attention should be given to the dilution and count statistics.*

NOTE: Either of the two scan test methods or an alternative method may be used for filter Types C, D, F, and G if agreed.

EN 1822:1998 has methods for testing filter medium and for testing filters in a test apparatus or in-situ. For testing filters in a test duct or in-situ, upstream and downstream sampling is done using a photometer, a device that uses scattered light intensity to determine penetration for filters classified H10 - H13, and a particle counter for filters classified H14 - U17. See chart below. In all cases, the downstream filter face is scanned with a probe to locate possible leaks.

Test reports are linked to the numbering on the HEPA/ULPA filter and show the flow during measurement, the pressure drop at the test volume flow, the collection efficiency measured for MPPS and the filter class derived from these parameters.[4]

Filter Class	Efficiency (%)
H 10	85
H 11	95
H 12	99.5
H 13	99.95
H 14	99.995
U 15	99.9995
U 16	99.99995
U 17	99.999995
H = HEPA U = ULPA	

Fire Resistance

The ability of a HEPA or ULPA filter to resist the effects of a fire is of major concern to users of these products. Usually, they are used in facilities intended to protect delicate products from external contamination or to isolate a hazardous product. The nuclear power and weapons industry is very much concerned about the fire resistance of HEPA filters used to capture any airborne nuclear materials. To establish safety criteria for HEPA filters used in such applications, Underwriters Laboratories developed a specific standard. This is the Standard for High Efficiency Particulate Air Filter Units, UL 586.

UL 586 (HEPA Filters)

HEPA filters can be listed as either Class 1 or 2 under UL 900 (See Chapter 7: *"HVAC Air Filter Testing"*), but this only shows the fire retardance of the filter and provides no data about its performance. There are situations when information about performance of HEPA filters after exposure to heat and flame is important. This is especially true for installations at processing plants for nuclear materials and for nuclear power stations. To meet these needs, UL 586[5] was established.

This standard describes the different materials of construction for HEPA filters and attaches only the following restrictions:

> If the frame material is wood it shall be treated to reduce combustibility by pressure impregnation and shall not be less than 3/4 thick.

The filter medium shall not contain unbonded asbestos fiber materials.

To conduct a UL 586 test a total of 15 filters is needed. All filters shall be representative of other sizes. Of these, 6 filters are tested in accordance with Section FC of the ASME AG-1 Code on Nuclear and Gas Treatment. This test is run in a penetrometer using Emery 3004 as the test aerosol source. The efficiency must not be less than 99.97% and the pressure drop across the filters at tested rating cannot exceed 1.3 in. w.g. (324 PA) for units rated at 25, 50, 125, 1500 and 2000 cfm and not more than 1.0 in. w.g. (249 Pa) for units rated at 500 and 1000 cfm. The disposition of these filters is not noted.

Aerosol Penetration Test

Six additional samples, sent to UL are subjected to the Aerosol Penetration Test in which an aerosol portable generator is used to generate the test aerosol.

The concentration of aerosol to be generated and the methods for zeroing the photometer and setting the upstream concentration at 100% are not spelled out. The only detail is that the penetration is to be measured downstream of the unit by means of a forward light-scattering cell and a percent penetration indicator. The percent of penetration through a filter unit shall not exceed 0.05%. The filters then are subject to additional testing.

Heated Air Test

Three of the filters that passed the aerosol penetration test are each individually installed in the test duct and subjected to a flow of heated air for 5 minutes. The temperature of the heated air must be 700 ± 50°F (371 ± 10°C) as measured at each of six points distributed across the filter face. The rate of flow measured by the calibrated venturi flow meter which is part of the test setup, must not be less than 40% of the manufacturer's rated airflow. After samples are cooled, they are again subjected to a UL aerosol penetration test. The DOP penetration shall be no more than 3%.

Moist Air Test

The fourth filter is stored in a static atmosphere with a relative humidity of 90 ± 5% at a temperature of approximately 77°F (25°C). After 24 hours, the filter is returned to the test duct and air passed through it at manufacturer's rated airflow capacity until the filter is dry. It is then subjected to the UL aerosol penetration test. The penetration cannot increase more than 0.01% above the as received value and cannot exceed 0.05%.

Low Temperature Test

Another of the six samples is conditioned for 24 hours in a static atmosphere with a relative humidity of 50 ± 5% at a room temperature of approximately 77°F (25°C). It is then transferred to a static atmosphere of 26.6 ± 4°F (-3 ± 2°C) for an additional 24 hours. It is allowed to warm up to room temperature and its UL penetration is again measured. The penetration shall not increase by more than 0.01% above the as received value and shall not exceed 0.05%.

Spot Flame Test

The final filter is subjected to a spot flame test that is intended to determine the fire resistance of the filter. A complete description can be found in the UL publication. The criterion for passing the test is that the downstream face of the filter shall not continue to flame for more than 2 seconds after removal of the test flame.

Confirmation Test

The final three filters shall be sent to either the Oak Ridge Filter Test Facility at Oak Ridge TN or to the Rocky Flats Test Facility at Golden CO where they are tested for pressure drop and penetration. Presumably these filters would be tested by the manufacturer before being sent to the test facility, but this is not stated as being required.

General

Filters meeting the requirement of UL 586 carry a distinctive UL label to distinguish it from the UL 900 label.

Notes to Chapter 8

1. Hinds, W et al. Size Distribution of "Hot DOP" Aerosol Particles Produced by ATI Q Aerosol Generator, 15th DOE Air Cleaning Conference, Boston (1978)
2. IEST-RP-CC001.4 HEPA and ULPA Filters, Institute of Environmental Sciences and Technology, 940 E. Northwest Highway, Mt. Prospect IL 60056
3. IES-RP-CC007.1 Testing ULPA Filters, Institute of Environmental Sciences and Technology, Mt. Prospect. IL
4. European Committee for Standardization NEN, Paques B. V., C. Molenaar 2004/0611
5. "Standard for High Efficiency Particulate Air Units," ANSI/UL 586-1990, Seventh edition (Oct. 1990) (rev. April 2000) Underwriters Laboratories Inc., 333 Pfingsten Rd., Northbrook IL 60062.

9

Controlled Environments

Introduction

Clean air is important for most production processes (see also Chapter 15: *"Industrial Contaminant Air Filtration Control Systems"*). The degree of cleanliness required will vary with the type of operation and the product involved

Almost all industrial processes that require clean air to protect a product depend upon some variation of the contamination control principle:

1. Clean the air to the extent required by the product or process being protected.
2. Blanket the product or process with cleaned air.
3. Use enough clean air to keep contaminated air away from the process.

Only the first step relates to the cleaning of air. The second and third relate to the airflow pattern and quantity of cleaned air. Filters can be selected to achieve the desired air cleanliness level, but if the air is not distributed properly or if the quantity of filtered air is inadequate, a product or process can still become contaminated.

Product and Air Cleanliness

If a product or process is to remain uncontaminated, it must be protected from all sources of contamination. These are: the air which surrounds the product or process, the container which holds the product, the operator who runs the process or handles the product, and the contamination which a process can cause to a product or vice versa. Cleaning the air and using it correctly solves only one part of the contamination control procedure. People may expect too much of an air filtration system if channels of contamination other than the surrounding air are ignored.

Cleanrooms

Federal Standard 209E, cancelled in November of 2001, describes a cleanroom as "a room in which the concentration of airborne particles is controlled and which contains one or more clean zones." FS 209 was replaced by the International Organization for Standardization (ISO) Standard 14644.

ISO defines a cleanroom as a:

> "...room in which the concentration of airborne particles is controlled and which is constructed and used in a manner to minimize the introduction, generation and retention of particles inside the room, and in which the relevant parameters, e.g. temperature, humidity and pressure are controlled as necessary."[1]

ISO defines a Clean Zone as a:

> "...dedicated space in which the concentration of airborne particles is controlled and which is constructed and used in a manner to minimize introduction of, generation and retention of particles inside the zone and which other relevant parameters e.g. temperature, humidity and pressure are controlled. ***NOTE: this zone many be open or enclosed and may or may not be located within a cleanroom."

Cleanrooms are recognized by their type and identified by their airborne particulate cleanliness levels. There are two main types of cleanrooms, depending on the method of air distribution within the room.

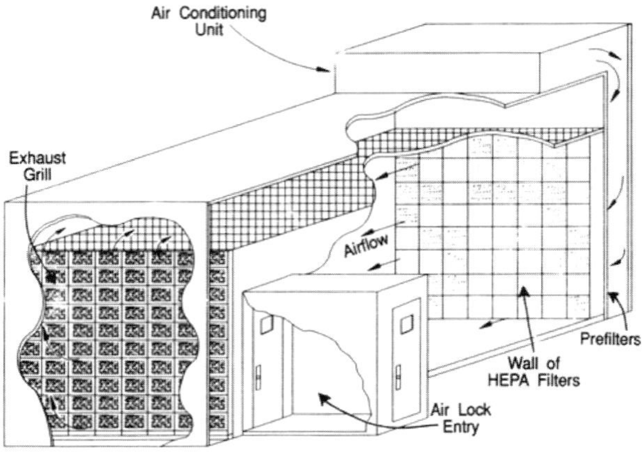

Figure 9.2. Old Style Horizontal Laminar Flow Cleanroom

Cleanroom Types

Non-unidirectional Airflow - These were previously known as non-laminar flow type and were originally referred to as conventional cleanrooms. In rooms like these the filtered air is usually introduced through non-aspirating diffusers. This air blankets the space below the diffusers (Figures 9.1 and 9.2) and mixes with the internal air by means of induction. Because of the discharge velocity of the air at the diffuser, the flow does not have parallel streamlines and contaminated room air mixes with the clean air. The mixture is removed from the room through multiple or continuous outlet grilles located at or near floor level in the case of vertical flow, cleanroomsat the oppositee wall end of a horizontal flow cleanroom. The intent, however, is still to surround the part or process with sufficient clean air to prevent airborne contamination. This clean air can be 100% outdoor air, but is usually a mixture of outdoor air and return air.

Unidirectional Airflow – with these types of cleanrooms, more commonly known as laminar-flow clean-rooms, the design intent is that air should make a single pass through the room with a piston-like effect. In one configuration, sometimes referred to as vertical laminar flow rooms, cleaned air enters the room through the ceiling and exits through floor grilles or through continuous outlets in the walls at floor level (Figure 9.3). When returns are in the walls, there are restrictions as to how wide the room can be and still maintain unidirectional flow.

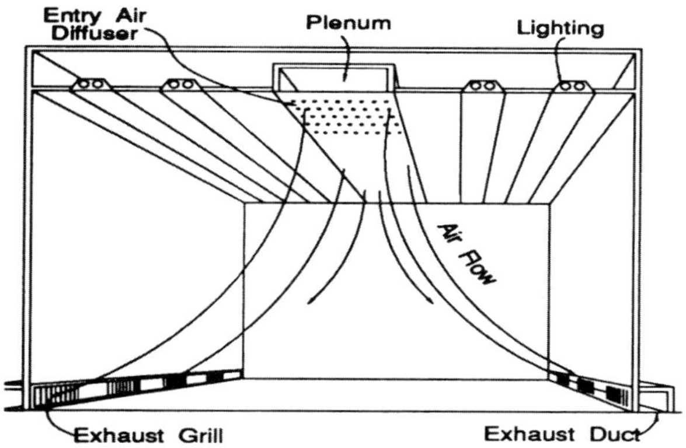

Figure 9.1. Conventional Cleanroom Continuous Air Diffuser

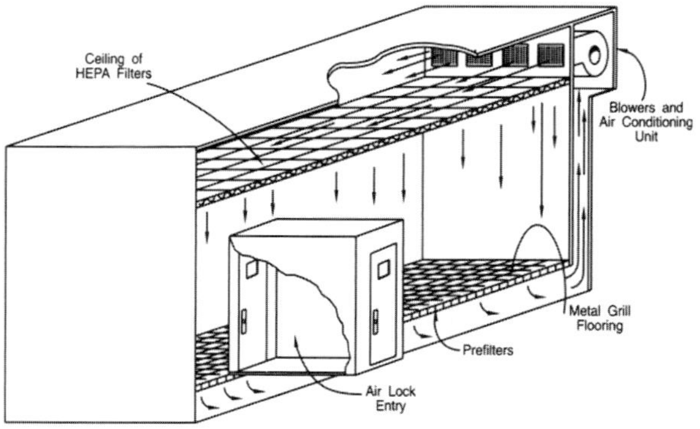

Figure 9.3. Modern Laminar Downflow Cleanroom with Grated Floor - Piston flow effect of air movement in room

In a conventional cleanroom, HEPA filters are usually located in the mechanical room and are the last items in the mechanical equipment sequence. They are downstream of the blower, so that only HEPA-filtered air enters the duct-work connecting the mechanical equipment to the cleanroom. The ductwork is frequently stainless steel, although other non-shedding materials are also used.

In a laminar flow cleanroom, the HEPA filters are located in the wall or ceiling of the room, so that once the air passes through the filter it is in the cleanroom. Intermediate sources of contamination are eliminated. HEPA filters must be sealed into the framing system so that there can be no by-pass of dirty air around them and into the cleanroom. The filters themselves must also be leak-free. Different filter manufacturers have developed framing systems that provide such support (Figure 9.4).

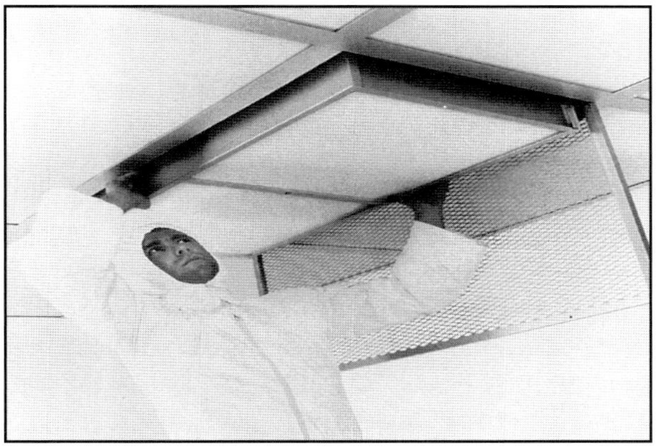

Figure 9.4. Installing HEPA Filter in a Framing System

Originally, the HEPA filters used in laminar flow cleanrooms were 5-7/8 in. (149 mm) deep. They used corrugated aluminum separators. This style filter is rated at a face velocity of 150 fpm (0.76 m/sec). Because they only operated at 100 fpm (0.51 m/sec) or less when used in a laminar flow cleanroom, their initial pressure drop was approximately 0.6 in. w.g. (0.373 Pa). Now many of the filters used for this application are the mini-pleated type with glue-bead string separators, or a separatorless design, operating at the same or lower pressure drop and having a depth of only 2 to 3 in. (51 to 76 mm).

HEPA or ULPA filters used in cleanrooms have a multiple year life expectancy, and usually require no prefiltration of the return air. However, the make-up of outside air for a cleanroom usually requires filtration and, depending on the cleanliness of this outside air, may use a MERV 6 or better prefilter and a MERV 14 secondary filter.

Downflow modules (Figure 9.5) are used for most cleanroom ceiling applications. These modules consist of assemblies in which HEPA filters are either factory sealed into housing modules making them a disposable module, or are replaceable HEPA or ULPA filters in a housing that is permanent. The housing modules, in turn, can be connected by flexible ductwork or other means to a supply air manifold. The advantage of a module is ease of installation, local control of face velocity using a built-in damper system, and flexibility to relocate the module to different parts of the room if manufacturing changes dictate such a change. Some newer designs incorporate individual blowers for each module and are frequently referred to as low-profile fan units (LP-FUs). This design is especially suitable for small isolated cleanrooms located in a larger space. If the air in the larger space has the required temperature and humidity, it is only a matter of cleaning the air by passing it through a HEPA or ULPA filter before discharging it into the cleanroom.

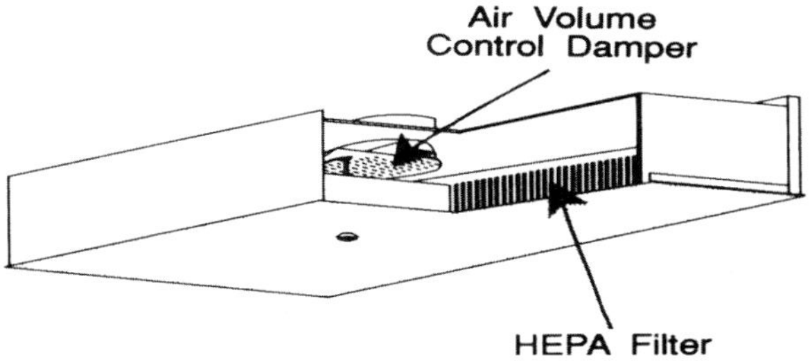

Figure 9.5. Downflow Module

First air or first position is used to describe the operation or product location which is first bathed by the HEPA filtered air in a laminar flow cleanroom. In a horizontal flow room, first air or first position would be that closest to the wall with the bank of HEPA filters in it. This air may pick up contamination as it makes its pass

Figure 9.6. Airborne Particulate Cleanliness Levels

	ISO/TC209 14644-1 Number of particulates per cubic meter by micrometer (µm) size					
Class	**0.1 µm**	**0.2 µm**	**0.3 µm**	**0.5 µm**	**1 µm**	**5 µm**
ISO 1	10	2				
ISO 2	100	24	10	4		
ISO 3	1000	237	102	35	8	
ISO 4	10,000	2,370	1,020	352	83	
ISO 5	100,000	23,700	10,200	3,520	832	29
ISO 6	1,000,000	237,000	102,000	35,200	8,320	293
ISO 7				352,000	83,200	2,930
ISO 8				3,520,000	832,000	29,300
ISO 9				35,200,000	8,320,000	293,000

over the first position. It may then be used to protect a second operation requiring less cleaned air. After becoming continuously more contaminated, the air exits the room through a wall opposite the one with the HEPA filters. In most downflow cleanrooms, first position is the only position. The air passes over the product or process to be protected and then leaves the room through the grated floor or low returns in the wall.

Cleanroom Classification

Cleanrooms are classified according to their level of cleanliness. In earlier versions of standards such as the now obsolete Federal Standard 209, there were only three classes designated by the maximum allowable number of particles 0.5 µm and larger in a cubic foot of air. These were Class 100, 10,000 and 100,000. In a later version there were six different classes based on I-P (English) measurements and 13 classes based on SI measurements.

ISO 14644-1 Classification of Air Cleanliness

This document is now an official ISO standard. It is the first of eight documents concerned with contamination control. The Institute of Environmental Sciences and Technology (IEST) is the secretariat for Technical Committee 209 of the International Standards Organization (ISO) charged with writing these standards. This standard is expected to replace all other existing ones.

Cleanliness is expressed in terms of an ISO Number for the level of cleanliness required for various size particles 0.1 µm to 5.0 µm. The standard specifically states that particles larger and smaller than this size range are not part of the standard. Only the SI (metric) system of measurement is used. The cleanliness classes are shown in Figure 9.6.

Downflow cleanrooms have become so popular that variations of them are used not only for the cleanest rooms, but also for those facilities where greater concentrations of dust particles are acceptable. Obviously, such rooms do not require the same degree of air cleaning or ventilation rate. However, it has been shown to be cost-effective to use HEPA filters but to use less of them to reduce the ventilation rate.

Several sources suggest the following percent of a ceiling that should incorporate HEPA filters in order to achieve the designated cleanliness levels:

Class ISO Class	Percent (Ceiling Area)*
5	90
6	75
7	60
8	40

Air cleanliness of less than ISO 5 requires the use of ULPA filters in the ceiling.

It should be kept in mind that these values are only generalizations and that there are a variety of factors that may affect them.

Clean Benches

There are situations when an operation must be performed under contaminant controlled situations but a cleanroom is not required. To meet this need, clean benches were developed.

Clean benches are enclosed workstations in which an operation or a product is bathed by air that has passed through a HEPA filter. They are supplied with their own blower and HEPA filter. In the horizontal flow version (Figure 9.7) the filter covers the whole face of the work station. It is important that the HEPA filter used in a clean bench is leak-free and that there is no leakage of contaminated air around the HEPA filter and into the work area. Clean benches with HEPA filters will usually provide ISO 3 conditions. Clean benches usually have prefilters at their air inlet.

Biological Safety Cabinets

Biological Safety Cabinets are a ventilated cabinet for personnel, product and environmental protection having an open front with inward airflow for personnel protection, downward HEPA filtered laminar airflow for product protection, and HEPA filtered exhaust air for environmental protection. Biological Safety Cabinets are used primarily for working with naturally occurring or exotic agents with a potential for respiratory transmission which may cause serious or potentially lethal infection. They are discussed more fully in Chapter 10: "*Airborne Microorganisms.*"

Replacement HEPA filters

Replacement of HEPA filters in clean benches and biological safety cabinets requires special precautions, education and testing equipment. They also may have specially manufactured HEPA filters. Therefore, when HEPA filters must be changed, special attention must be taken, preferably requesting the manufacturer and part number (or equivalent), so that if there are any special modifications necessary for the filter to fit the framing system or clean bench, they are clearly identified.

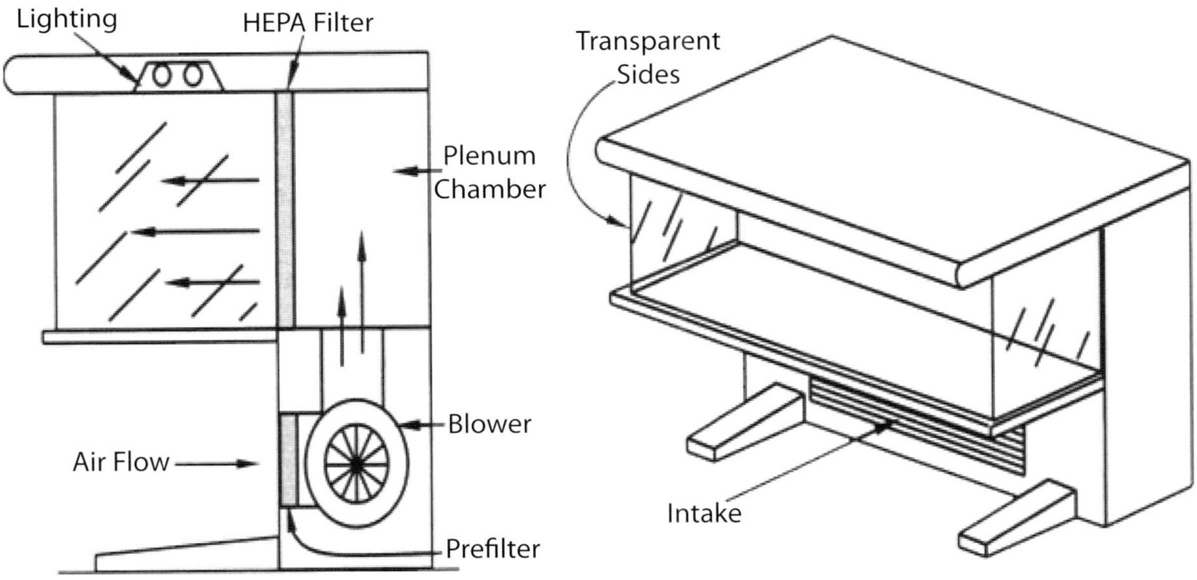

Figure 9.7. Horizontal Flow Clean Bench

Industrial Finishing

Industrial finishing consists of the various operations and activities intended to give the surface of a product the characteristics and color it needs for its intended use and appearance. The next sections deal specifically with the finishing process.

It is an oversimplification but convenient to consider that industrial finishing occurs in a box. There are several different box configurations, but in all of them cleansed air is introduced and contaminated air is drawn out of the box. This is referred to as intake and exhaust air respectively. When air exhausted exceeds air that is introduced, provisions must be made to replace the air that has been exhausted. This is referred to as makeup air. Sometimes an additional amount of makeup air is supplied to cause outward directional flow from the box.

Exhaust Air - The first requirement for ventilation in most finishing processes is in the exhaust from the area where the finishing is taking place. It has many purposes. In spray painting it is to control the overspray, the spray material that does not coat the item being painted. Without ventilation, overspray would soon escape to contaminate the surroundings and create potential health, fire and safety hazards in the area where the spraying is done. An exhaust fan draws air through the space where spraying is being done, capturing the overspray and pulling it away from the painter and the parts being painted, then through some type of air filter before it is discharged, usually to the outdoor atmosphere. Another important purpose of ventilation is to remove paint solvent vapors from the spray area to protect workers from volatile organic compounds (VOCs) exposure and prevent the build up of solvent vapor to a potentially explosive level.

Exhaust ventilation is required for other processes such as dipping, flow coating and roller coating to remove solvent vapor during the drying phase. Ventilation speeds drying by creating a condition favorable to having the solvent evaporate more quickly so that the paint dries faster. Likewise, ventilation is required to prevent solvent vapors from escaping into the work area where they can become a health hazard to workers.

Supply Air - Makeup air is required to replace the air that has been exhausted. Frequently the temperature and humidity of the air must be controlled to provide the environment most suitable for the finishing process. Equally important, the air must be cleaned so that it will not damage the finish of the item being coated. Sometimes the dedicated supply air is fed directly into the finishing area. Air distribution systems help set up air flow patterns to provide the most effective ventilation of the finishing area. More sophisticated systems are discussed in page 9.9, "Finishing Rooms."

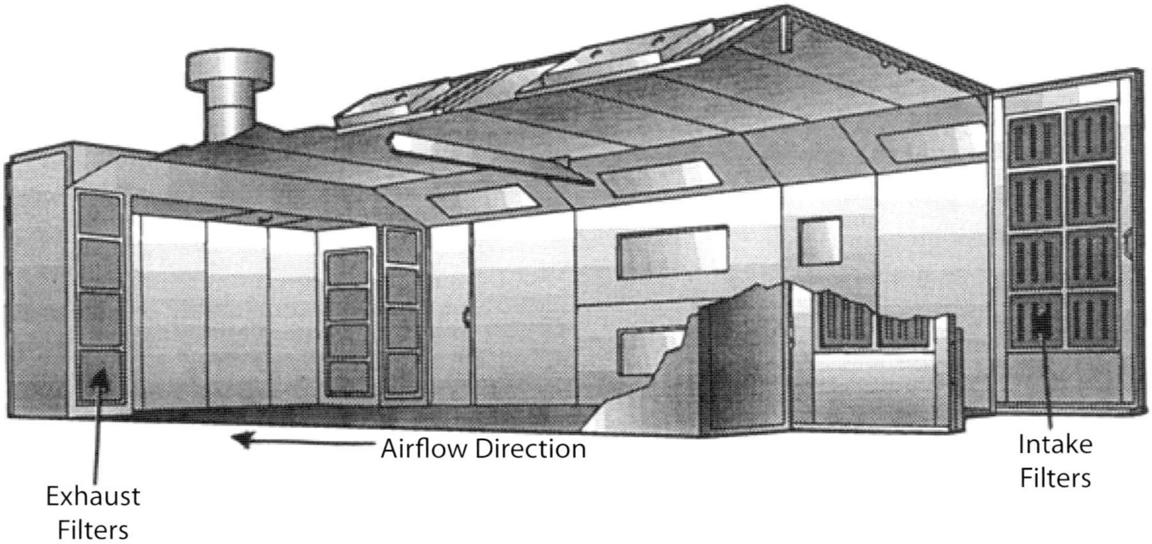

Illustration Courtesy Airguard Industries

Figure 9.8. Crossdraft Spray Booth

There are other situations where finishing is accomplished in a room that, for convenience, has openings in its walls. The dedicated makeup air system supplies filtered and tempered air to the area surrounding the finishing operation. The exhaust system is designed to move the surrounding air into the finishing area, through the air cleaner, and then to discharge it to the atmosphere.

Other times, the makeup air comes from the building area surrounding the finishing operation. If there are circumstances where the building air contains objectionable dust, this dust must be removed before it can be used as makeup air. One solution to the problem is to enclose or seal off the operation area while it is in use, putting filters in the doors and walls to clean the makeup air before it enters the finishing area. The filters should have a MERV of 6 or greater (Figure 9.8). Because ANSI/ASHRAE Standard 52.2 standards indicates efficiencies based on particle size, by knowing what size particle is unacceptable in the finish of the work, a filter can be selected that eliminates that size particle from the air transferred to the finishing area. Care must be taken to assure that the filter system (frames, and other openings) are sealed to maintain the system integrity and not allow air bypass.

Relationship between Supply and Exhaust Air

It is a basic principle that supply air volume must equal (and possibly slightly exceed) exhaust air volume. This slight amount of excess supply air causes a positive pressure between the finishing area and the area surrounding it. This positive pressure may be required to be certain that unfiltered air does not enter the finishing area through cracks or other openings in the barrier between the finishing area and the rest of the building.

Not all supply air needs to be delivered directly to the finishing area. In those instances where finishing areas involve the spray-painting of coatings with EPA-listed hazardous components, a negative pressure (frequently enough to maintain a design inward flow of 150 fpm)[2] may have to be maintained in the finishing area to prevent the escape of hazardous materials into the surrounding area. In this case the supply air would be less than the exhaust air to allow for the negative pressure.

Desired control of a quality finishing operation is rarely achieved if an exhaust system is used alone without a separate matching supply air system.

Balancing Supply and Exhaust Systems

When the supply air and exhaust systems for a finishing operation are first started, it is possible to manually balance the systems so that airflow is as desired. However, as the systems operate both will change. The supply system filters will load with environmental dust and the paint arrestors with overspray, resulting in an increase in resistance. Both increases will affect the respective air quantities handled by the supply and exhaust fans. It is expected that the paint arrestor filters will load up more rapidly than the supply filters and will have to be changed more often.

To keep the systems in balance, two pieces of equipment are required:

A sensor must recognize that there is an unbalance. This can be:

1. A sensor must recognize that there is an unbalance. This can be:
    a. A hand-held measurement device such as a velometer or anemometer.
    b. An air flow measurement device in both the supply and exhaust.
    c. A sensitive pressure transducer that can recognize that the proper balance between exhaust and supply systems is being maintained.
2. Air flow controllers to adjust the systems so that they are in balance. These controllers get necessary information from the sensors. They can be:

a. Volume control dampers in both the supply and exhaust systems.

b. Inlet vane dampers (for centrifugal fans)

c. Variable frequency drives (VFD) to change the speed of the fan motors as needed.

The choice of a control system depends on many variables including size of the air handling systems and the control system's owning and operating costs. However, some form of monitoring is needed to be sure that the supply and exhaust systems are in the desired balance and that desired airflow velocity is maintained.

Exhaust Air Cleaning

The exhaust from a finishing operation usually contains overspray (liquid, wet solid or dry materials) and solvent vapor (gas). Removal of overspray from the exhaust system is accomplished by the installation of suitable collectors at the exit of the finishing area before the exhaust air enters the ductwork leading to the exhaust fan and is discharged to the outdoor atmosphere. The most frequently used collection devices are air filters and air washers. The primary purposes of either of these items are:

1. To capture the overspray so that it can be easily disposed.

2. To minimize accumulations in exhaust ducts to prevent the possibility of fire hazard.

3. To remove the material so that it will not accumulate on fan blades and reduce fan performance.

4. To prevent the discharge of nuisance or toxic quantities of overspray into the atmosphere.

Removal of solvents is also an important consideration. Most solvents are volatile organic compounds (VOCs) that are considered environmental and/or safety hazards. Regulations require that the ventilation rate be high enough that the concentration of a VOC vapor is held below 25% of its lower explosive limit. The environmental hazard arises from the discharge of these vapors into the atmosphere where they contribute to air pollution. Solvent removal is accomplished by either adsorption or incineration. Adsorption is the capture of the solvent vapors by adsorbents such as activated carbon and zeolite. Some forms of adsorbent media can be reactivated.

Another alternative is to incinerate the waste. This can be done by catalytic destruction or by burning in an incineration device. In some systems the energy of this combustion can be recovered. When the VOC is a pure hydrocarbon, the by-products of incineration are carbon dioxide and water. However, if the VOC contains chlorine, sulfur and some other acid-forming elements, the by-products can include hydrogen chloride and sulfur dioxide, both of which may have to be removed from the exhaust gas before it can be discharged into the atmosphere.

Readers are referred to Chapter 11: *"Airborne Molecular Contaminants,"* for more information on adsorption but are reminded that the concentrations of VOCs are significantly lower in the normal environment than in the exhaust from finishing. For this reason, gas, gas-phase or molecular filters designed for normal HVAC systems may have limited use in finishing exhaust systems.

Supply Air Filtration Systems

The level of air cleaning required for an air supply system will depend primarily on the quality of finish required on the painted product. It is conceivable that for some rough coated operations, no filtration is required. On the other hand there are luxury items and quality merchandise where no surface blemish is permitted. Tests have shown that particles as small as 5 μm are responsible for visible blemishes on smooth surfaces such as automobiles with a clear-coat finish. If this is the case, the makeup air must contain no particles 5 μm and larger. Currently the selection of a filter suitable to achieving this level of cleanliness is a matter of experience, however makeup air systems will usually include filters up to MERV 14. If the finishing operation is in a building in a dirty industrial area, prefilters may be used to extend the life of the final filter. These prefilters are usually MERV 6 or higher. In the case of continuous finishing operations e.g. automobile bodies, filter users frequently want the time between air filter changes in the make-up air system to match the model changeover of the product being painted or at preplanned plant shutdowns. In order to accomplish this, more filters or filters of greater capacity are used to extend the time between changes.

Finishing Room or Paint Spray Booth

There is always the possibility that particulate matter may flake off from ductwork downstream of the supply filter system or find its way into the cleaned supply air system. To prevent these secondary particles from damaging finishes, diffusion media (media that is installed at the point where clean air enters the finishing room or booth) is frequently used in paint spray booths. Newer design spray booths function as vertical laminar flow rooms with the diffusion media as the room ceiling. The diffusion media is intended to remove all particles 5 μm or larger that may be in the supply air. It also diffuses the air to create uniform airflow over the finishing operation. Diffusion media is usually made from fire-retardant synthetic fibers, formed into a mat that is highly resistant to fiber shedding. The media is frequently treated with a tackifier that prevents migration of any large particles that may be captured. Rather than installing the media in pads or ring panels as a separate framing system that may allow by-pass, paint spray booths are frequently designed so that the media can be unrolled as one complete blanket or a series of blankets to cover the spray booth ceiling (Figure 9.9). Because it reflects light and allows for a better rendition of the color of the painted item, white is the preferred color for diffusion media.

In a situation where makeup air is not fed directly to the finishing room but instead is sent to the area surrounding the room, the filtration requirements are the same as when makeup air is sent to the room directly.

Crossdraft paint spray booths are finishing areas where makeup air is part of the plant HVAC system. When appearance is not critical, the booth may be wide open at one end. When finish appearance becomes more important, and there is excessive contamination in the plant itself, filters with a MERV 6 or higher are used to clean the makeup air. These are typically located in the access doors (Figure 9.8).

Types of Finishes

In addition to VOC-based paints, water-based, powder coating and ultraviolet light UV-curable paints are becoming more popular. The

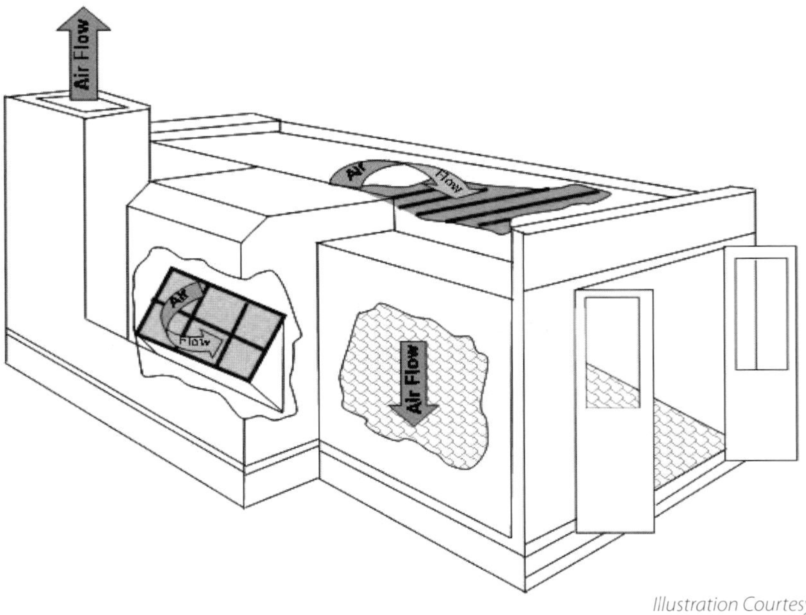

Illustration Courtesy Airflow Technology, Inc.

Figure 9.9. Downdraft Spray Booth

substitution of water for a significant portion of the volatile organic solvents provides two important considerations. First is the cost saving in using water instead of VOCs. The second is the reduction in the need for equipment to prevent the escape of VOC vapors into the atmosphere through the finishing exhaust system.

The most popular growth area in the development of finishing products has been in the use of dry spray materials (powder coatings). Instead of suspending pigments and dissolving them in solvents, dry powders containing all the components are used. Generally they are used in electrostatic processes where the object to be finished is charged opposite to the dry spray or is grounded. In either case the dry spray is attracted to the item to be finished where it is deposited on the suitably prepared surface. The coating can then be thermally cured and become permanently attached to the item being painted. Other curing methods are available.

One unique consideration in the use of dry coatings is the ability to recover and reuse over-spray. The exhaust is not pulled through paint arrestors. Instead it is drawn into a recovery unit where the overspray is removed. In one style the air is passed through high efficiency cyclone separators where much of the larger over-spray is recovered. The air then passes through special reverse flow high efficiency filters in which the smaller particles collect. Reverse flow keeps the pressure drop of the filters at a low state, and the filters themselves are designed so that the collected material can fall out of them and be returned to the recovery system.

Photo Courtesy Columbus Industries

Figure 9.10. Paint Spray Booth Exhaust Filter Media

Some of the recovery systems are so efficient, the exhaust air can be recirculated back into the finishing area if the original powder is free of any hazardous materials. The power requirements for recovery systems is much higher than for an exhaust system since the pressure drop of the cyclones and final filters combined may exceed 5 in. w.g. (1243 Pa).

Ultraviolet curable finishes is a radiation curing technology that utilizes electromagnetic radiation energy to effect chemical and physical changes of organic finish materials by the formation of cross-linked polymer networks. One type of radiation is UV light. The UV curable process represents an approximately 83% reduction in VOCs per weight of solids applied. Some examples of products using UV curing include automotive headlights, tail lights, reflectors, and wheel hubs, and also in the coating of wood furniture.

Paint Arrestor (Filter) Categories

Currently there are a variety of designs of paint arrestors and filter media used in the capture of overspray. These include:

Expanded Paper consists of multiple layers of Kraft paper that has been slit and expanded. The finished product is approximately 1 in. (2.54 cm) thick. It is available in rolls, batts and pads. (Figure 9.10)

Fiberglass Media is a progressive density spun glass material produced primarily in 1 in. to 4 in. (2.54 to 10.16 cm) thickness. It is available in rolls, batts and pads.

Both Expanded Paper and Fiberglass Media are available with a downstream layer of synthetic media. The addition of the synthetic layer is claimed to increase the overspray removal and paint holding capacity.

Synthetic Media is a multiple denier synthetic bonded fiber produced in a progressive structure. The product can be multiple denier and multiple density in construction to increase progressive density. Synthetic media are used in many combinations (composites) to adjust air flow and meet different efficiency requirements. Polyester media is available in both dry form and with tackifier. The media is primarily produced in 1 in. to 2 in. (2.54 to 5.08 cm) thickness and is available in rolls, batts, pads and extended media filters. Some manufacturers are now making a convoluted polyester filter that has the appearance of an egg crate. The added surface area is claimed to have higher paint-holding capacity than flat polyester media of the same polyester compound.

Baffle Type Filters use primarily the impingement principle of particle removal. They are currently available in several designs and materials that include Kraft board, cardboard, Styrofoam and metal. Air passes through the media fabricated to have venturi type openings in it. These openings cause the air to accelerate, increasing the velocity of the air and overspray so that as the air passes through other openings the large over-spray particles strike and stick to the backup layer. A downstream layer of synthetic media may be used to capture the finer particles in the airstream.

Other filters, usually associated with HVAC systems with efficiencies up to and including HEPA filters may be used for very special conditions. In any case, where high efficiency filters are used, they are preceded by one or more lower efficiency filters to remove most of the overspray and leave the final filter to collect the small amount of very fine particles that pass through the prefilters.

At the present time there is not enough standardized information available for choosing filter media and construction for any specific paint arrestor application. For instance, in addition to filter test report data, such details as media type, product construction, percent of solids in the paint and wetness of overspray should be taken into consideration in selecting a filter system.

NAFA distributors or filter manufacturers should be contacted for their recommendations for the preferred type of media to be used for any specific paint arrestor application.

Water-Wash Systems

For a long time water-wash systems have been used to capture overspray. A typical design consists of a recirculating pump which takes water from a sump and distributes it evenly over the face of a surface creating a "curtain" called a weir to capture the larger overspray. The exhaust air itself is directed toward and through a sinuous collector element at water level in the sump. In this process much of the finer overspray is captured.

The overspray is collected in the sump as a sludge that is removed during routine maintenance. Problems with disposal of the sludge and the additional power needed to pump water and pull the exhaust air through the collector element have caused this system to lose favor to other types of arresting filter systems.

Water-wash systems have not been found to be effective in the removal of VOCs.

Special Applications

Currently, some applications fall under the guidelines of the National Emissions Standards for Hazardous Air Pollutants (NESHAP). At this time the aerospace industry has had emission standards published. This includes manufacturing and rework facilities. Existing facilities require a two stage system and new facilities require a three stage system. Independent testing must establish efficiencies in both the liquid and solid phase that match or exceed the requirements published. Testing is conducted under EPA 319 guidelines and is similar to the ANSI/ASHRAE Standard 52.2 tests. (*See Appendix Six.*) As NESHAP establishes emission standards for additional industries these types of systems will become more prevalent.

Disposal of Overspray Pads and Filters

Paint-laden overspray filters may also be a disposal problem depending on regulation of waste. The Federal Resource Conservation and Recovery Act ((RCRA) pronounced *Ric-Ra*) is directed so the "...waste are managed in an environmentally sound manner."

The basic tenet of this regulation is a "cradle-to-grave tracking system" meaning that hazardous waste generators and users of hazardous materials who generate hazardous wastes must track waste from the moment it enters the site as a hazardous material to the eventual treatment or disposal of that material. For further information on this topic go to www.epa.gov/rcraonline. Check regulations for disposal guidelines and determination of the class of the waste.

Care should always be taken in the storage and disposal of used overspray media and filters. Rolled up or stacked media pads have been know to be susceptible to spontaneous combustion when stored or left in an enclosed dumpster.

Other Typical Controlled Environments

(See also Chapter 15, "*Industrial Contaminant Air Filtration Control Systems*").

Firing Ranges

Commercial firing ranges present a control and removal problem for lead dust from bullets and fumes to which the shooter and employees of the range are exposed. The use of controlled airflow in a laminar direction combined with HEPA filtration/molecular filtration, and dilution ventilation with outdoor air can provide a safe and effective means of protection. For complete information, consult the "*NAFA Guidelines for Firing Ranges,*" available at www.nafahq.org.

Libraries, Archives and Museums

Preserving important and priceless historical documents, artifacts and paintings requires a high level of particulate and molecular filtration. Important and irreplaceable artifacts have been lost by exposure to sulfur dioxide, nitrogen dioxide and ozone.

Viable particles in the form of airborne fungal spores, bacteria and mold can deteriorate artifacts. Non-viable particles cause soiling where cleaning of the artifacts is not an option. Prefiltering with MERV 8 or higher filters, installation of molecular filters and final filters of MERV 15 or higher is considered the best way to preserve these artifacts. For complete information, consult the "*NAFA Guidelines for Libraries, Archives and Museums,*" available at www.nafahq.org.

Nuclear Energy Applications

Air filters, especially HEPA filters, are important elements in providing safeguards for the handling of nuclear products. This includes the processing of materials to make nuclear fuel, the peaceful use of fuel in nuclear power plants, and the processing of spent fuel.

In nuclear power plants, these filters are used in control room emergency air supply systems and on the exhaust from containment vessels where the nuclear reactors are installed. They have been an equally important means of controlling radioactive exhaust where materials for nuclear weapons have been processed or manufactured.

Because of the nature of the application, air filters used in exhaust systems from nuclear applications are subject to stringent quality control considerations. Specifications may require that the HEPA filter be on the Department of Defense (DOD) Qualified Product List (QPL) or that the filters must be tested at a Department of Energy (DOE) filter test site. There is also an plethora of regulations which may apply. For this reason, anyone supplying replacement filters for these applications should carefully examine the specifications to be certain that all requirements are met.

Computer Equipment

The size and the amount of heat generated by computer equipment has been reduced significantly over the years. Along with this has come a corresponding reduction in the amount of air conditioning required. In rooms where there are larger computer consoles, the equipment is frequently installed on raised floors which provide a runway for cables and a plenum into which conditioned air is supplied. This is called an underfloor air distribution system. Natural convection of air warmed by the electronic equipment assists in drawing the conditioned air up through the cabinets, where it cools the equipment. Returns are at the ceiling. Filters with a MERV 9 and higher are normally recommended for this application to prevent dust from accumulating on the solid-state circuits. If outdoor air contains unusually high quantities of dust, higher-efficiency filters may be necessary to clean this air before it is introduced into the computer room.[3] A filter efficiency of MERV 13 meets most requirements.

Photographic Industry

Quality photographic products must be free of any dust particles that will impair the quality of the final image. This is especially true of high-resolution film which may be enlarged by thousands of magnifications. Even a small dust particle on the film may end up as a large black spot on an enlargement. To prevent such contamination, HEPA filters are used extensively to filter the air over the coating and drying processes. In most instances the HEPA filters are part of the air conditioning system for the coating rooms. Much of the coating process is proprietary for different manufacturers.

The medical industry has been especially affected by microscopic particles of dust. For example, a small speck of dust on a mammogram will create a spot that could be misinterpreted as a possible tumor. In very critical photography such as high altitude aerial films and for microminiature images, the use of laminar airflow rooms or workbenches is recommended.

Breather Filters for Storage Tanks

As a tank of liquid is filled, air inside the tank must be allowed to escape. Likewise, as the tank is emptied, air enters the tank to replace the liquid. If the liquid must be protected from contamination, provisions must be made to be certain that the air which moves in and out of the tank is clean.

In many food and pharmaceutical processes, the tanks are sealed but have a vent port through which the air can flow. An appropriately enclosed filter can be attached directly to the vent port. HEPA and molecular filters are available in sealed units ready for such direct attachment. The rising or falling liquid in the tank acts as a piston moving air through the filter.

Sometimes tanks may require permanent access openings for the addition of material or for the withdrawal of samples. In these instances, a self-contained filter-blower unit can be connected to the tank. This unit should have a capacity adequate to have outward airflow through all openings of at least 100 fpm (.508 m/sec) under all conditions of tank use.

Industrial Process Control Rooms

In many industrial processes, the whole operation is controlled electronically by computers located in a control room. Operators monitor the information generated by the computers and modify instructions as necessary to create the desired product. This type of operation is likely to be found in the processing of mineral ores, petrochemical refineries, paper mills, and in chemical process buildings.

Most control rooms are located in areas where there are airborne chemicals, both particulate and gaseous, that are corrosive and if allowed into the control room, would attack computer circuit boards and other exposed contacts, either destroying them or causing them to short circuit. To prevent this from happening, filtered air must be introduced into the room to maintain outward flow (positive pressure) through all openings between the room and its surroundings. This air must be free of all chemical contaminants, both particulate and molecular, and large deep-bed carbon or other suitable molecular filtration is used. As an extra precaution, HEPA filters are often used to capture particulate material. Depending on the nature of the airborne particles, this may not be necessary, but the cost and importance of control equipment frequently dictates using filters of the highest efficiency.

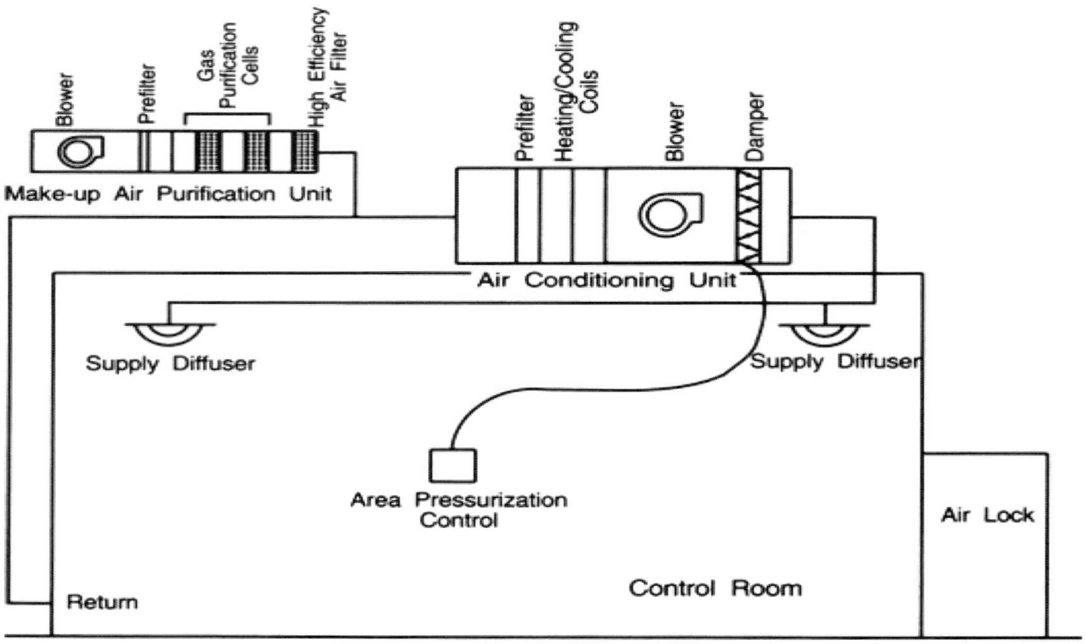

Illustration Courtesy Airguard Industries

Figure 9.12. Industrial Control Room Ventilation

The removal of the gaseous components requires molecular filters. These filters are discussed in detail in Chapter 11: "*Airborne Molecular Contaminants.*"

Control rooms are usually air conditioned, so there is a flow of air from the room, through conditioning equipment, and then back to the room. Since outside air is required for the occupants of the control room, it must also be cleaned. The air needed for pressurization may be adequate to meet the outside air requirement for the room occupants. If it is not, part of the return air can be vented to the atmosphere so that it can be replaced by the required makeup air. The pressure drop through the outside air purification system may represent a high portion of the overall system resistance while the outside air is only a small part of the air being handled. For this reason, it makes sense to add a separate blower system to supply the cleaned outdoor air to the air conditioning system which can then control temperature and humidity as needed for the room (Figure 9.12).

Sewage Disposal Plants

Most large sewage disposal plants use the activated sludge method to convert sewage into a disposable sludge. In this process sewage is introduced into a tank along with "seed" sludge. Large quantities of fine bubbles of air are passed through this mixture causing the bacteria in the seed sludge to multiply by feeding on the sewage material. The end product is a fine floc which settles to the bottom of the tanks where drag conveyors move it up and out of the tank. Part of this sludge is recycled as "seed sludge" and the balance is sent to liquid filters where it is dewatered and dried, or sent to a digester where it is allowed to decompose further, producing methane gas which is used for power at the sewage disposal plant.

The success of a sewage disposal system depends on dissolving as much oxygen as possible in the activated sludge tank water to encourage the growth of bacteria that convert sewage into sludge. To do this, very fine bubblers of different designs are placed in the bottom of the tank. Because they are so fine, the bubblers are easily clogged. Consequently, all particles which could cause clogging must be removed from the air before it passes through them. In order to accomplish this level of cleaning, MERV 14 filters have proven satisfactory. The specific filters selected to reach this level may vary. They are usually prefiltered by filters MERV 6 or higher.

The activated sludge process, when in proper operation, is essentially odorless, principally because the process is aerobic, i.e., there is an adequate supply of oxygen to meet all the biological demands. When the sludge is dewatered, and if it is digested, anaerobic bacteria take over. They thrive in an absence of oxygen. The anaerobic process takes place in a sealed tank and results in the production of gases, one of which is methane as has been mentioned previously. In addition there is a wide spectrum of gases, many of which are sulfur-based. These gases are highly corrosive. Material exposed to them must be made out of stainless steel or coated with a corrosion-resistant material. It is also necessary to remove these gases before air containing them is vented to the atmosphere. They must also be removed from the plant ventilation system. To accomplish this a suitable molecular filter must be used.

Textiles

Textile plants take in baled cotton, wool, or synthetic fibers and ship finished fabrics. During this process, a variety of operations take place depending on the raw material and the finished product. For instance, in the processing of cotton "white goods" (bedding materials of different types) cotton is carded to align the fibers that are formed into a sliver, a relatively thick cord of low density. This sliver is processed through a variety of machines that cause it to end up as a thread. This thread may then be dyed. After this, it is treated with sizing to give it body during the weaving operation which follows. After weaving, the fiber may again be washed and dried.

There are large amounts of dust generated when cotton bales are opened and the cotton is carded. These machines have their own exhaust system, which when properly operated, allows little dust or lint to escape. The major portion of airborne dust in a textile plant is lint from the various operations involved in turning the sliver into a thread and the sizing dust generated during the weaving operation. Byssinosis, (pronounced bis-a-no-sis) also known as "brown lung" is a respiratory hazard in mills where unprocessed cotton is handled.

The control of temperature and humidity is very important for the operation of a textile plant. Cleanliness levels have also been set by OSHA specifying the maximum concentration of dust to which a worker may be exposed.

Humidification is most frequently accomplished by various forms of air washers. In addition to helping control dust and lint within the textile mill, air filters help air washers. If filtration is inadequate, lint accumulates on the air washers impeding the flow of both air and water. It may also clog the water recirculation system of the washer. Sizing dust, composed in part of starch-like compounds, is captured by the washer, coating its surfaces with a gel-like material. This material ends up in the washer water, where the gel can clog washer sprays and generally make maintenance a nuisance. In those cases where there is inadequate filtration, washers may require shut-down maintenance as frequently as twice a week.

Unitary extended-surface filters incorporating synthetic fibers have been used successfully for this application. They are usually underrated and depend principally on straining as the filter mechanism. The lint coat which forms on the filter acts as a filter aid to capture the sizing dust. Filters are not changed when they are loaded but rather are vacuum cleaned to remove the accumulated lint and sizing. Eventually the lint and sizing particles migrate into the media, causing a filter loading which requires a filter change.

Where dust and lint loads are heavy, special rotary self-cleaning filters have become the units of choice.

Gas Turbines

Gas turbines are used for many stationary power applications, especially for generation of electricity. When a utility faces peak demands, turbo-generators can be quickly brought on-line to meet this demand and can be taken off as the demand declines. Meanwhile, the steam power stations can operate at their constant peak efficiency, filling the needs of the base load. Gas turbines operate on essentially the same principle as aircraft jet engines. Air is taken into the inlet and compressed by passing through a series of rotating vanes. The compressed air is mixed with gas and the mixture burned under controlled conditions. This hot mixture expands, turning more rotors which transfer some of the power back to the compressor, but most of the power is used to turn direct-connected electric power generators. In contrast, most of the energy from an aircraft jet engine is converted into thrust which pushes the aircraft. The exhaust from the gas turbine passes through a heat exchanger to produce steam that can be used for heating and industrial processes.

Photo Courtesy Airflow Products Co., Inc.

Figure 9.11. Gas Turbine Filter

Airborne dust creates two problems for gas turbines. Particles larger than 10 µm can cause erosion of turbine compressor blades. Particles 5 µm and smaller cause deposits on and fouling of turbine blades.[4] Erosion causes permanent damage, reducing the ability of the compressor section to adequately compress the air. Smaller size particles have the same effect. By building up on the turbine blades they affect the blade configuration and consequently its air handling characteristics.

The preferred filter for urban industrialized areas is a MERV 14. (See Figure 9.11). In other areas where the air is cleaner, MERV 12 filters may be adequate. Where there is a heavy load of particles greater than 10 µm, inertial separators are sometimes used. These too take various configurations but in general the dust particles are moved out of the airstream by their inertia as the airstream abruptly changes direction. Ten percent of the air entering inertial separators is drawn off into a secondary bleed air circuit incorporating its own blower. This circuit conveys the large particles from the separator either to the atmosphere or to a separate dust collector.

Telecommunication Equipment Rooms

The rapid change which has occurred in telecommunication is perhaps no better illustrated than by the fact that facilities once known as "telephone exchanges" are now rightly called telecommunication equipment facilities. When electromechanical and analog switches used with rotary dial systems were the primary equipment within these facilities, the major effects of airborne dust were associated with large particles. Efficiency in the range of MERV 4 to 6 filters were used in rural, suburban, and small city locations. In larger cities and industrialized areas, filters of MERV 13 and 14 were used to prevent the damaging effects of small particles. These effects included metallic corrosion (stress cracking) and conductive bridging of electrical components. Molecular filters, such as activated carbon, have been and are presently still being used in locations with local corrosive or highly polluted ambient environments.

The advent of digital technology has led to much smaller and much more densely packed equipment. Increasingly, smaller diameter (including submicron) particles now affect the operation of this equipment. An increasing emphasis has been placed on air filtration as the primary control mechanism for the reduction of dust-related equipment failures. A large percentage of telecommunication equipment facilities now utilize high-efficiency air filtration in both large and small air conditioning systems. Depending upon the location, size, occupancy, and other factors, MERV 12, MERV 13 or MERV 14 filters are normally used. With the increase in legislation meant to control ambient air pollution particularly through the Clean Air Act, it is expected that the use of molecular filtration equipment in telecommunication facilities will decrease or remain approximately the same. A possible exception to this is the option of using molecular filtration to reduce outdoor air requirements for buildings with high internal generation rates of volatile organic compounds, or for industrial sites with very strong local ambient sources, such as paper mills and petrochemical refineries.[5]

Other Applications

For each industrial application named here, there are many more that have not been referenced. Air cleaning problems associated with mechanical equipment have been solved by incorporating filters into the equipment at the time of manufacture. An example of this is a cleanroom vacuum cleaner which comes with a built-in HEPA filter. This filter is located so that it will not only remove all the fine dust passing through the vacuum cleaner bag, but also will capture the carbon dust created by the wear of the carbon brushes used in the vacuum cleaner motor.

For the majority of applications, filters are used on supply or recirculated air to capture any dust in the ambient air. There are some occasions when filters are used on exhaust system, however, except for nuclear and biological applications where HEPA filters are a must, caution should be observed in using any filter on an exhaust system connected to a dust collection hood of any type. Air filters are not suitable for applications in which there are heavy concentrations of dust [in excess of 4.36 grains per 1000 cu ft (10 mg/m3) because they will quickly load and require frequent changes.

Notes to Chapter 9

1. ISO 14644 Part 1 and Part 3, International Organization for Standardization, 2002.
2. 29 CFR.1926.57 (h)(6))(I) OSHA Regulation "Ventilation"
3. ASHRAE 2003 Applications Handbook. Chapter 17: Data Processing System Areas, page 17.4.
4. Dennis A. Nordberg, Air Filters -Types, Selection and Application (Additional information not available)
5. This material derived from information provided by Mark E. Krzyzanowski of Bellcore.

10

Airborne Microorganisms

Introduction

The air around us contains a variety of microorganisms from many sources including bacteria, fungal spores and viruses. Airborne bacteria enter a milk container and can cause the milk to sour. As people sneeze, cough and talk, they generate thousands of fine droplets of moisture, many of which contain organisms that become airborne.

- Talk - 3,000 Droplets
- Cough - 4,500+ Droplets
- Sneeze - 40,000 Droplets

When the water in the droplets evaporates, they become droplet nuclei, solid particles, most of which can stay airborne indefinitely. Various forms of legionella bacteria thrive in ponds and in untreated cooling tower water. They can become airborne in sprays from cooling towers or even from the wind stirring up pond water and are responsible for Legionnaire's Disease.

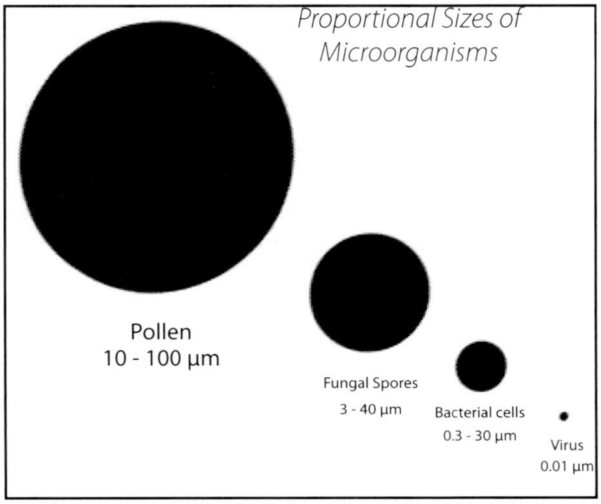

Airborne microorganisms can be captured by sampling the air using a variety of devices. Individual organisms cannot be seen except under a microscope. However if they are allowed to grow in a suitable nutrient, they multiply rapidly forming colonies which can be seen by the naked eye. Each microorganism or cluster of microorganisms which grows to form a colony is called a colony-forming unit (CFU).

The term microorganism includes all organisms that can be found in nature. In this category by order of size, smallest to largest, they are:

- **Viruses** - metabolically inert, viruses are an infectious agent essentially of a core of RNA or DNA surrounded by a protein coat. Viruses replicate only within the cells of living hosts. They multiply by entering a host cell and using the cell's DNA to manufacture clones of themselves. In this process the host cell DNA is destroyed, its walls ruptured, and the newly formed viruses are released. They move on to other hosts and repeat the cycle. Examples in humans are the common cold, influenza, chickenpox, smallpox, measles, and mumps.

- **Bacteria** - single-cell organisms which are spherical, spiral, or rod-shaped and appear singly or in chains and may be free-living, saprophytic, or pathogenic in plants or animals. All multiply by single cell division. Examples of diseases in humans caused by bacteria are tuberculosis, pneumonia, rheumatic fever, whooping cough, botulism, and even tooth decay.

- **Fungi** - any of a diverse group of eukaryotic single-celled or multinucleate organisms that live by decomposing and absorbing the organic material in which they grow, comprising mushrooms, molds, mildews, smuts, rusts, and yeasts. They reproduce through the formation of spores that separate from the parent organism. Under suitable conditions, these spores attach themselves to a food source and begin the growth of a new colony. Certain molds are characterized by their ability to survive and grow under conditions where other forms of microorganisms would either be dormant or die. The persistent survival of this class of microbes is illustrated by the mildew and other mold that appear in a variety of damp conditions, on cheese and other foods, on water wash cooling coils, and in any location where even the slightest nutrient is available.

There are many other paths by which microbes can be transported from one location to another. However, our concern is for airborne microbes.

A naked airborne microbe is relatively rare (fungal spores are an exception). Instead, it may be attached to a dust particle, surrounded by dried fluids (droplet nuclei), or may exist in clumps with other microbes. This fact can distort filter efficiency test data, because one cannot always be sure if it is a naked microbe that is being captured or if it is only one part of a droplet nucleus. A colony-forming unit can be one microbe or it can be a cluster of them.

Filter Efficiency

Most tests to determine the efficiency of different filters on airborne microorganisms have been performed using microbes that are not harmful to humans. Two bacteria have been used that meet this requirement. They are:

- *Serratia indicia*, a bacterium that is 1 μm long and 0.5 micron in diameter.
- *Bacillus subtilis varniger* spores that have a diameter of 1 micron. Because this is the spore form of a bacterium, it is very stable and not easily damaged.

In those instances where the airborne microbe of specific concern is hazardous to any other living organism, tests are run under controlled conditions so that this microbe does not escape into the atmosphere.

There are two viruses that have been used extensively. They are:

- E. coli T3 phage that has a particle size range of 0.02 to 0.05 μm.

- T1 phage that has a diameter of 0.1 micron

Both of these phage are harmless to humans but attack bacteria.

The filter testing process involves growing a batch of microbes. They are isolated from the growth media as much as possible and then used to make a stabilized broth containing a known concentration. The broth is nebulized (made into a fine spray) in the test airstream. Samples are captured upstream and downstream of the filter, and cultured in suitable nutrient on a Petri dish. In the case of bacteria, each cell will grow and produce a colony that can be counted. In the case of viruses, bacteria susceptible to attack by the virus are grown on Petri dishes so that there is a uniform layer of them. Whenever a virus contacts this bacterial layer, it will attack the bacteria, destroy them, and leave a plaque (hole) in the bacterial culture. Each plaque is considered to be derived from a single virus. The plaque can be counted and the concentration of the viruses calculated.

Removal of Microorganisms from the Airstream Using HEPA Filters

Viruses The same reports that provided information on the removal of bacteria (see below) also provide data about the removal of viruses. The report[4] which gave details on the bacterial removal efficiency of commercially available HEPA filters also included the results of tests run on these same filters using T1 phage as a challenge. The results of these tests are as follows:

Filter	DOP Penetration (Percent)	T1 Phage Penetration (Percent)
1	0.012	0.0039
2	0.020	0.00085
3	0.0060	0.0044
4	0.0020	0.0010
Average Penetration		0.0025%
Average Efficiency		99.9975%

Bacteria The use of HEPA filters to remove bacteria from the airstream has been well documented. One report[1] gave the result of tests on four different HEPA filters that had been purchased on the open market. In this series of tests, the Bacillus subtilis spores mentioned previously were used as a challenge. The summary of the results of four tests on each filter were as follows:

Filter	DOP Penetration (Percent)	T1 Phage Penetration (Percent)
1	0.012	0.00011
2	0.020	0.000072
3	0.006	0.00028
4	0.002	0.0023
Average Penetration		0.00069%
Average Efficiency		99.9993%

Other tests confirm these values. Allen[2] tested HEPA filter media using a challenge of Staphylococcus aureus. This organism is less than 1 micron in diameter. He found that the efficiency on the test aerosol was 99.97%.

Removal of Microorganisms from the Airstream Using HVAC Filters

Filters incorporating microfine fibers are now used in many applications where the control of airborne microbes is a concern.

In 2003, NAFA participated in a research project completed by Research Triangle Institute under a contract with the United States Environmental Protection Agency entitled, "Filtration Efficiency of Bioaerosols in HVAC Systems." The purpose of the project was to determine the ability of standard HVAC filters to remove potentially harmful bacteria and viruses from an airstream in one pass.

Three bioaerosols were used in the testing:

- The spore form of the bacteria *Bacillus atrophaeus* (BG) , a gram-positive spore-forming bacteria elliptically shaped with dimensions of 0.7 to 0.8 by 1.0 to 1.5 μm
- *Serratia marcescens,* a rod-shaped gram-negative bacteria with a size of 0.5 to 0.8 by 0.9 to 2.0 μm
- The bacterial virus, (bacteriophage) MS2 dispersed as a micrometer-sized polydispersed aerosol.

Filters were tested in MERV categories from 6 to 16, and in both the clean and loaded configuration. Results were as follows:

	Removal Efficiency		
Challenge	B. atrophaeus	S. marcescens	MS2 phage
MERV 16	%	%	%
Clean	99.4	99.5	99.3
Dust Loaded	99.7	99.8	99.6
MERV 15			
Clean	99	99	95
Dust Loaded	99	99	99
MERV 14			
Clean	89	83	87
Dust Loaded	99	99	99
MERV 12			
Clean	64	74	70
Dust Loaded	88	92	93
MERV 8			
Clean	54	58	57
Dust Loaded	74	83	79
MERV 6			
Clean	25	45	40
Dust Loaded	83	88	91

Environmental Technology Verification, Research Triangle Institute, 2003, David S. Ensor, et. al.

Removal of Microorganisms from the Airstream Using ESP's.

In 1951, researchers at Ft. Detrick determined the efficiency of electronic air cleaners on an airborne bacterial challenge.[3] Using *Serratia indicia* as the challenge, they tested two electronic air cleaners. The one designated Model E was the older version. Model F included improvements which would increase the charge intensity on particles in the ionizing section and so increase the probability that they would be captured in the collecting section. Test results were as follows:

Model E	92% efficient
Model F	97% efficient

From this testing, they concluded that, "Although electrostatic precipitators can remove a high percentage of bacteria and dust from the air, they may not be satisfactory as filters where a constant supply of clean air is required. Electrostatic precipitators that receive maximum maintenance have been shown in laboratory tests to remove or destroy approximately 90 percent of the microorganisms in the air. However, tests of some units under normal operating conditions have shown much lower efficiencies. Under optimum physical conditions and with satisfactory maintenance, electrostatic precipitators can be used in place of medium efficiency filters. They should be equipped with high-or ultra-high-efficiency filters downstream if the air is to be supplied to critical areas. Without maximum maintenance, electrostatic precipitators can give a false sense of security."[4]

Removal and Retention of Mold

Microbial filter tests have shown that removal is size related and suggest that the efficiency on mold spores, which are the same size as bacteria, will be at least as high as that on bacteria. Mold has been known to grow on the organic material captured by high-efficiency filters. What is of concern is the possibility that mold might grow through the media and that spores might be liberated on the downstream side. In the case of HEPA filters, there is the controlling fact that proprietary fungicides are added to media made to meet the mildew resistance requirements of the military specification MIL-F-51079.

Antimicrobial Agents

There has always been concern that filters, when loaded, could become a suitable breeding ground for bacteria and mold. The proposed solution has been to add an antimicrobial coating to filter media fibers to prevent such growth from occurring.

There has been much conflicting evidence on this point. For this reason ASHRAE funded a research project entitled "Determine the Efficacy of Antimicrobial Coatings of Fibrous Air Filters," (ASHRAE Research Project RP 909, June 2000). The research was done using 3 types of antimicrobial coatings:

1. Quaternary Ammonium Compound
2. An iodophor
3. An amine neutralized phosphoric acid ester

The report answered some of the questions about antimicrobial coatings and raised others. In short, field-tested filters showed that filters that incorporate an antimicrobial coating, may still be susceptible to microbiological growth when loaded. When more extreme testing was done in the lab, microbial growth was dependent on the type of antimicrobial coating used.[24]

Air Washing

"Washing alone is not a satisfactory method of removing bacteria from air since the efficiency is usually relatively low. Air washers tested have been found to remove 20-80 percent of bacteria in the 1-5 micron range. In some instances where wash water is recirculated, the bacterial count of the air may actually increase because of re-aerosolization of bacteria that have accumulated in the water. Air washers cannot be recommended for the removal of bacteria from air supplied to critical areas."[5]

Disease Control

Microbes are responsible for various diseases which attack not only humans, but also plants and animals. With regard to humans, it was noted:

"One method of transmission of disease-producing organisms is through the air (airborne infection). An aerosol is a suspension of particles in air, in this particular instance, a suspension of microorganisms in air. In nature, organisms may get into the air from the wind blowing over water and soil, raising microscopic water particles and dust particles into the air and suspending them there. Potential pathogens get into the air of human environment from sneezing, coughing and talking (droplet nuclei) as well as from clothes, bedding, sweeping of floors, and other similar sources. Under such conditions these factors affect the spread of infection: (1) the organism's virulence; (2) the host susceptibility; (3) the effect of the environment on the organism either preventing or aiding in its survival."[6]

Healthcare Facilities

While healthcare facilities provide important services in helping cure the sick, they are, by their very nature, places where cross-contamination is likely to occur unless stringent environmental control measures are in place. One of the channels requiring control is that of airborne contamination. One method of control is air filtration.

The establishment of standards for air filtration for healthcare facilities rests with state health departments. The Hospital Survey and Construction Act, also known as the Hill-Burton Act, was a United States federal law passed in 1946. This act responded to the first of President Truman's proposals and was designed to provide federal grants and guaranteed loans to improve the physical plant of the nation's hospital system. Included in these standards were requirements for air filters.

The Hill-Burton Act was amended in 1975 and became Title XVI of the Public Heath Service Act (see www.fda.gov/opacom/laws/phsvcact/phsvcact.htm). The American Institute of Architects, Academy of Architecture for Health (with assistance from the U.S. Department of Health and Human Services) has assumed responsibility for the publishing of the AIA Guidelines for Design and Construction of Hospital and Healthcare Facilities.[7]

Listing of requirements for filter efficiencies for healthcare central HVAC systems is shown in Figure 10.1. In the past, most of the emphasis was on general hospitals. Now, guidelines include healthcare facilities such as psychiatric hospitals, nursing care facilities and outpatient facilities.

Additional requirements emphasize the importance of proper installation. They include the following:

1. Filter holding frames shall be durable.
2. Filter holding frames shall be sized to provide an airtight fit within the enclosing ductwork.
3. All joints between filter frame banks and the enclosing ductwork shall have gaskets or seals to provide a positive seal against air leakage.
4. A filter pressure measuring device shall be installed across each filter bed having a required efficiency of 75% (MERV 12) or more, including hoods requiring HEPA filters. Provisions shall be made for access to pressure resistance devices for field testing (calibration).
5. Non-central AHU's shall be equipped with a permanent (cleanable) or replaceable filter with a minimum MERV 3 (68%) average ASHRAE test dust arrestance).

In addition to the different levels of filtration, there are many variables for the use of the filtered air in different areas. These include: (1) air movement relationships to adjacent areas, (2) minimum changes of outside air per hour, (3) minimum total air changes per hour, (4) recirculation by means of room units, and (5) need for all air to be exhausted directly outdoors.

Specialized Operating Rooms

The 2007 ASHRAE Applications Handbook[7] includes a recommendation for orthopedic operating rooms, bone marrow transplant rooms, and organ transplant rooms. This recommendations includes MERV 8 filters for the first filter bank and MERV 14 for the second bank with the third bank consisting of terminal HEPA filters at each outlet to the room.

Figure 10.1

Filter Efficiencies for Healthcare Central Ventilation And Air Conditioning Systems

Published by the American Institute of Architects, Academy of Architecture for Health and Assistance from the U.S. Department of Health and Human Services

Area Designation	Minimum No. of Filter Beds	Minimum Filter Efficiency [1]	
		Filter Bed No. 1 MERV (%)	Filter Bed No. 2 MERV (%)
General Hospitals			
All areas for inpatient care, treatment and diagnosis and those areas providing direct service or clean supplies such as sterile and clean processing, etc	2	8 (30%)	14 (90%)
Protective environment room	2	8 (30%)	17 (99.97%)
Laboratories		13 (80%)	-
Administrative, bulk storage, soiled holding areas, food preparation areas and laundries	1	8 (30%)	-
Recirculation of air within Individual Isolation Rooms	1	99.97% DOP (HEPA)	-
Psychiatric Hospitals			
All areas for In-Patient Care, Treatment and Diagnosis and those areas providing Direct Service.	2	8 (30%)	14 (90%)
Administrative, Bulk Storage, Soiled Holding, Laundries, Food Preparation areas.	1	8 (30%)	-
Nursing Care Facilities			
All areas for resident care, treatment, and/or diagnosis, and those areas providing direct service or clean supplies	2	7 (30%)	13 (80%)
Administrative, Bulk Storage, Soiled Holding, Laundries, Food Preparation areas.	1	7 (30%)	-
Outpatient Facilities			
All areas for inpatient care, treatment and diagnosis and those areas providing direct service or clean supplies such as sterile and clean processing, etc.	2	8 (30%)	14 (90%)
Laboratories	1	13 (80%)	-
Administrative, bulk storage, soiled holding areas, food preparation areas and laundries	1	8 (30%)	-

Notes

1) *Additional roughing or prefilters should be considered to reduce maintenance required for filters with efficiency higher than 75%.*
2) *Efficiency ratings shall be based on ANSI/ASHRAE Standard 52.1-1992 and shall be average. MERV based on ANSI/ASHRAE Standard 52.2-1999.*
3) *Non-central AHU Systems shall be equipped with permanent (cleanable) replaceable with a minimum MERV 3 (68% average weight arrestance efficiency.*
4) *Filter Bed Location. Where two beds are required, filter bed No. 1 shall be upstream of A/C components. Filter Bed No. 2 shall be located downstream of any fans or blowers (Blow-through System).*
5) *Humidification equipment shall be located 15' minimum upstream of Filter Bed No. 2.*
6) *Filter housing blank-off panels shall be permanently attached to frame constructed of rigid materials, and have sealing surfaces equal to or greater than filter efficiency installed in frame.*
7) *Filter measuring devices shall be installed across each filter bed having a required efficiency of 75% (MERV 12) or more, including hoods requiring HEPA filters.*

The use of terminal HEPA filters does not imply that these operating rooms would be laminar flow rooms. There are laminar flow operating rooms in use in many hospitals, however medical authorities are not in agreement that there is a need for them in all hospitals where specialized work is done.

Airborne Infection Isolation Rooms (AII)

The concept of Airborne Infectious Isolation (AII) rooms is not a new idea. The emergence of HIV and multidrug resistant strains of tuberculosis, have increased the importance of room pressurization, air change rates, filtration and air flow patterns in these rooms (Rousseau and Phodes 1993). The need for improving such facilities has worked in two directions. (1) Since they have no natural resistance to infectious organisms, immune suppressed people, i.e. HIV, need to be protected from airborne pathogens. (2) People with infectious tuberculosis (TB) need to be isolated from other patients, healthcare workers (HCW) and the rest of the hospital.

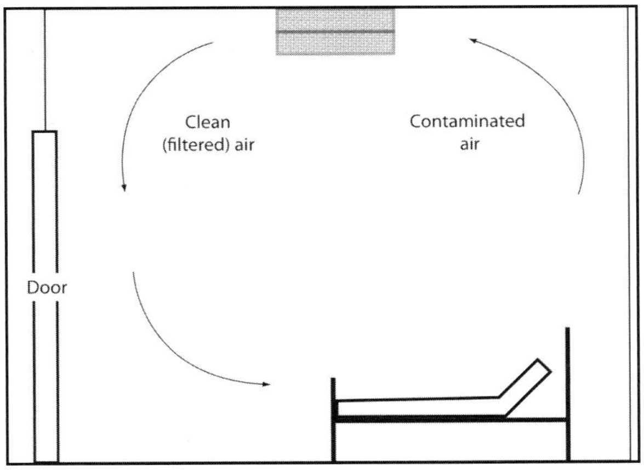

Courtesy AeroMed, Inc.

Figure 10.2. Example of in-room air cleaner in an isolation room

The emphasis is on quality facilities, engineered to prevent any airborne cross- contamination. In the case of the immune suppressed person, the patient room must be maintained under positive pressure to prevent the inward leakage of any contaminated air. In the case of the person with an airborne infectious disease such as TB the patient room must be under negative pressure so that there will not be any outward flow of air that may contain hazardous organisms. Isolation rooms with dual-purpose or switch-reversible airflow mechanisms that allow rooms to be switched between positive and negative pressure are no longer acceptable. Small rooms separating an isolation room from a corridor called anterooms are also recommended.

Tuberculosis Control

Because of the tubercle bacillus' ability to be easily transmitted person-to-person causing an increase in the number of cases in the United States, and with the discovery that the germ had become even more resistant to antibiotics, the Centers for Disease Control and Prevention produced "*Guidelines for Preventing the Transmission of Mycobacterium tuberculosis in Healthcare Settings*" in 1994. This document, updated in 2005, emphasized minimizing its spread through administrative and environmental controls: [9]

- Individuals with suspected or confirmed TB must be placed in an isolation room.
- The isolation room must be kept under a negative pressure of 0.01 in. w.g. or exhaust must be greater than the supply by at least 125 cfm.
- Isolation rooms must have more than 12 air changes per hour (ACH).
- Patients with active tuberculosis must have their rooms monitored with a smoke test daily to confirm that the room is under negative pressure.
- Ventilation can be by any of the following, all of which must include HEPA filtration:
    - Facility HVAC system
    - Fixed room air recirculating unit
    - Wall- or ceiling-mounted recirculating unit
    - Portable room air recirculating system
- A pressure-sensing device shall be installed to determine when the filter needs to be replaced.

- HEPA filter maintenance should be performed only by adequately trained personnel.
- Unoccupied rooms must be ventilated for an appropriate period of time after a patient leaves in order to achieve 99.9% removal efficiency before permitting employees to enter without proper respiratory protection. This would require 35 minutes @ 12 ACH.

Air-cleaning methods (e.g., room-air recirculation units containing HEPA filters or UVGI systems) can be used to increase the equivalent ACH.

Pathology and Bioresearch Laboratories

Biohazard is a combination of the words biological and hazard that represent infectious agents that present a potential risk to the health of plants, people, or animals, through direct infection or indirectly by the environment.

There are a variety of research laboratories in which microbes of varying virulence and chemical or viral carcinogens of different types are used, consequently, different levels of biological control for these facilities have been established. Biosafety levels are a combination of laboratory practices and techniques, safety equipment, and laboratory facilities appropriate for the operations performed and the hazard posed by the infectious agents and the laboratory function or activity.[21] There are four general descriptions:

Biosafety Level 1. This laboratory possesses no special engineering design features.

Biosafety Level 1 practices, safety equipment, and facilities are appropriate for undergraduate and secondary educational training and teaching laboratories. They are also appropriate for facilities in which work is done with defined and characterized strains of viable organisms not known to cause disease in healthy human adults. Biosafety Level 1 represents a basic level of containment that relies on standard microbiological practices with no special primary or secondary barriers recommended, other than a sink for hand washing.

Biosafety Level 2. Practices, equipment and facilities are applicable to clinical, diagnostic, teaching, and other facilities in which work is done with the broad spectrum of indigenous moderate-risk agents present in the community and associated with human disease of varying severity. With good microbiological techniques, these agents can be used safely in activities conducted on the open bench, provided the potential for producing aerosols is low. Hepatitis B virus, Human Immunodeficiency Virus, the salmonellae and Toxoplasma spp. are representative of microorganisms assigned to this containment level. Even though organisms routinely manipulated at Biosafety Level 2 are not known to be transmissible by the aerosol route, procedures with aerosols of high splash potential, which may increase the risk of such personnel exposure, must be conducted in primary containment equipment or in devices such as biosafety cabinets (BSC) or safety centrifuge cups.

Biosafety Level 3. Practices, safety equipment, and facility design and construction are applicable to clinical, diagnostic, teaching, research, or production facilities in which work is done with indigenous or exotic agents with a potential for respiratory transmission and which may cause serious and potentially lethal infection. Biosafety Level 3 places more emphasis on primary and secondary barriers to protect personnel in contiguous areas, the community, and the environment from exposure to potentially infectious aerosols. For example, all laboratory manipulations should be performed in a BSC or other enclosed equipment, such as a gas-tight aerosol generation chamber. Secondary barriers for this level include controlled access to the laboratory and ventilation requirements that minimize the release of infectious aerosols from the laboratory.

Biosafety Level 4. Practices, safety equipment, and facility design and construction are applicable for work with dangerous and exotic agents that have a high individual risk of life-threatening disease, which may be transmitted via the aerosol route and for which there is no available vaccine or therapy. The laboratory worker's complete isolation from aerosolized infectious materials is accomplished primarily by working in a Class III BSC or in a full-body, air supplied, positive pressure personnel suit. The Biosafety Level 4 facility itself is generally a separate building or completely isolated zone with complex, specialized ventilation requirements and waste management systems to prevent the release of viable agents to the environment.

Biological Safety Cabinets (BSC)

Biological safety cabinets are used extensively in microbiological research, work on hazardous microorganisms and the mixing of anti-cancer drugs. They are found in most hospital laboratories and pharmacies. They are categorized as follows:

Class I – ventilated cabinet for personnel and environmental protection, having an unrecirculated inward airflow away from the operator that exhausts all air to the atmosphere after filtration through a HEPA filter. Class I cabinets are suitable for work where no product protection is required. (Figure 10.3.)

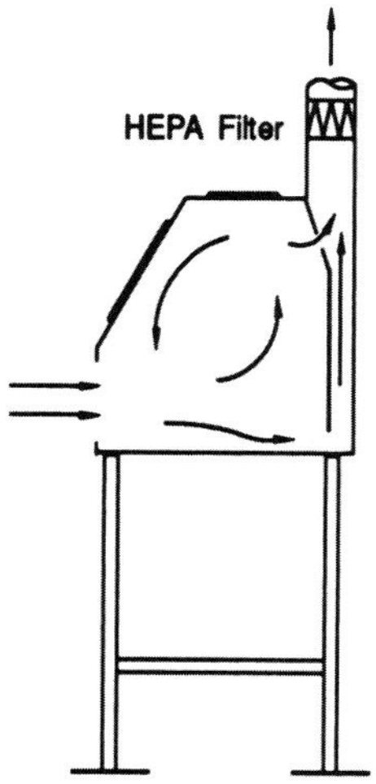

Figure 10.3. Cutaway View Showing Airflow at Work Opening, Class 1 BSC

Class II – A ventilated cabinet for personnel, product and environmental protection having an open front with inward airflow for personnel protection, downward HEPA filtered laminar airflow for product protection and HEPA filtered exhaust air for environmental protection. These cabinets come in Type A1, Type A2, Type B1, and Type B2 depending on the inflow and downflow velocities, ducted exhaust and recirculation design. (Figure 10.4).

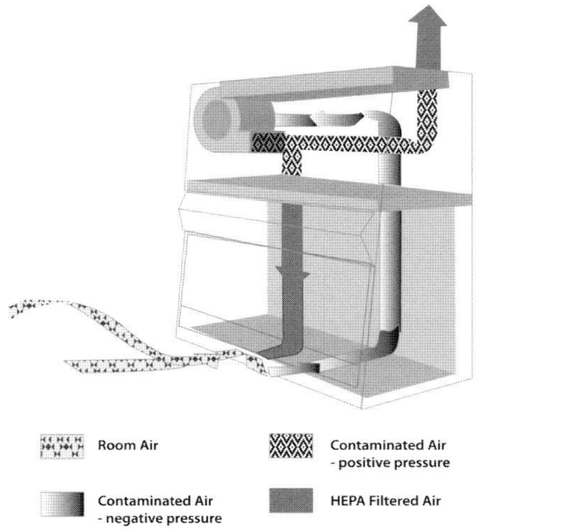

Courtesy Baker Company

Figure 10.4. Class II Biological Safety Cabinet

Class III – a totally enclosed, ventilated cabinet of leak-tight construction. Operation in the cabinet are conducted through attached rubber gloves. The cabinet is maintained under negative pressure of at least 0.50 inches w.g. (120 Pa). Downflow air is drawn into the cabinet through HEPA filters. The exhaust air is treated by double HEPA filtration or by HEPA filtration and incineration. [22] (Figure 10.5.)

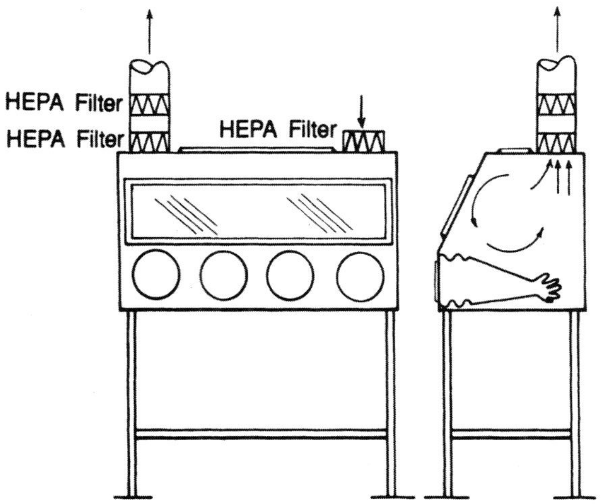

Figure 10.5. Class III, Biological Safety Cabinets

All biological safety cabinets require periodic certification and filter change. This should only be done by trained personnel experienced in the proper use of biological safety cabinets. Organizations such as the National Sanitation Foundation (NSF), have certification programs for technicians doing this type work.

Readers are referred to NSF Standard 49, *"Class II (laminar flow) biosafety cabinetry," Revised September 2004.* [23]

HEPA Filter Decontamination

HEPA filters that are used on biological safety cabinets or on the exhaust from research laboratories may become contaminated with pathogenic organisms or chemicals which may be highly carcinogenic.

If the contaminant is a pathogenic organism, it may be destroyed by gases such as formaldehyde or other members of the formaldehyde family. For this method of decontamination, gas-tight dampers are installed in the ductwork before and after the filter. In addition, provisions are made for means to introduce the decontaminating chemical and to allow decontamination to occur completely before the filter is removed.

One method which can be used for protection against any biological or chemical agents is the bag-in, bag-out procedure described in Chapter 5: *"HEPA, ULPA and Super ULPA Filters."* Once the contaminated filter has been successfully bagged, it can be removed to a decontamination facility where the destruction of pathogens or neutralization of chemical carcinogens can be performed.

UV lighting is not recommended in Class II biological safety cabinetry. If requested by the purchaser, it shall be installed in such a manner that it does not reduce the required performance.[25]

Food Processing

Many processes for the production of food products depend on microbial action. In most of these processes there is a three-phase program requiring different levels of air filtration and contamination control:

1. Microorganisms are cultured and evaluated in laboratories to determine the specific strain which has the best ability to produce the desired end product. Work on these cultures is usually performed in Class II biological safety cabinets to protect the strain from contamination.
2. Larger cultures are grown from these strains under controlled conditions. The level of control depends on the culture and product involved. The rooms in which this is done are relatively small and are designed for good housekeeping. Air sanitation can vary from the use of local control using MERV 14 filters.
3. The larger cultures are used to seed commercial batch runs of material to be converted into the end product. The actual production of the product will be performed in an environment that depends on the quality of the product. This is illustrated in some of the examples that follow.

Cultured Milk Products

Cultured milk products are those in which bacteria are carefully selected to grow in the milk, resulting in such products as cottage cheese, yogurt, and buttermilk. Since flavor is so important, there is a continuing selection and testing of bacteria cultures to improve the quality of the end product.

Bacterial cultures are grown in laboratories under controlled conditions that prevent the contamination by foreign bacteria, viruses that may attack the bacteria, and mold that may compete for food. Once the cultures have grown to the desired end point, they may be dried and aseptically packaged so that they cannot be contaminated. This work may be done in a biological safety cabinet, a horizontal laminar flow clean bench, or a cleanroom.

The dried culture is shipped to the production plants in whose laboratories it is reconstituted with sterile water. The reconstituted culture is then added to sterile nutrient and the container stored in a warm water bath to encourage the growth of bacteria. The resulting product is called the starter. Starter rooms are designed for easy sanitation. Desired air sanitation is accomplished by using MERV 14 air filters.

Starter is added to large tanks of milk that are kept at the desired temperature to produce the end product. In the case of cottage cheese and other cheese products, the culturing takes place in open tanks. When a proper cheese curd has formed, it is cut, washed to remove traces of whey, and then treated to get the desired characteristics of the end product.

Large quantities of warm to hot water are used in the production facilities. In addition there is overhead equipment of different types. The challenge is to prevent condensation from occurring on any of the overhanging devices and dropping into the open tanks of the product. The other part of the challenge is to prevent the growth of any bacteria or mold in whatever condensate occurs. To do this requires the use of air conditioning equipment that will minimize condensation. These same air conditioning units frequently include MERV 14 filters.

Milk Drying and Other Processes

Accepted practices for the handling of milk and milk products are described in "3-A" documents prepared by the three sponsoring organizations listed in Reference 13-15.

Milk drying is a process in which skim milk is sprayed into a hot airstream causing the moisture to evaporate and leaving behind a solid milk product—dried milk. A 3-A Accepted Practice[14,15] for this operation required that processing air which will be heated before product contact shall pass through a properly installed and maintained filter(s) which has a of MERV 14 when tested in accordance with the ANSI/ASHRAE Standard 52.2, when operating at its design face velocity. Processing air which will not be heated before product contact shall be passed through a properly installed and maintained filter(s) which has a minimum efficiency of MERV 13.

Instantizing is the process in which dried milk and other dry milk products are moistened, re-dried, and cooled in such a manner as to substantially improve their dispersing and re-liquefaction characteristics.[16,17] The air cleanliness requirements for this operation are the same as for milk drying.

Aseptic Packaging

Aseptic packaging is the process in which a product is packaged in a sealed container in an essentially sterile environment. The product may be naturally free of contamination, or it may be sterilized or pasteurized before packaging. One side of the challenge is to protect the product from the time it is made "sterile" until the time it is packaged. The other side is to protect the package from the time it is made sterile until it is filled and sealed.

In most processes all the work is done in enclosed equipment so that the product or the sterile package is not exposed to the atmosphere of the room in which the filling occurs. Many aseptic packaging units incorporate blower systems to keep the equipment under positive pressure. HEPA filters on the discharge of the blower essentially sanitize the air before it is used to pressurize the equipment. In equipment where access is required, the blower-filter capacity is selected which will provide outward flow of filtered air through all access openings while the equipment is in operation.

Pharmaceutical Industry

One of the first major applications for the HEPA filter was in the fermentation process by which penicillin was manufactured. The process required that compressed air be bubbled through the nutrient medium in order to encourage the growth of the organism producing the penicillin. If this air was not free of foreign organisms, they could also grow in the nutrient medium. This could result in a loss of yield because of the use of food intended for penicillin producing organisms or because of the production of materials which made the purification of penicillin more difficult.

A second problem relating to penicillin was the fact that many people were allergic to very small amounts of penicillin itself. It could not be allowed to become airborne and possibly contaminate other products. Manufacturing activities involving penicillin powder had to be totally isolated so that there was zero contamination of other products that might be made nearby.

To accomplish this, the manufacturing areas were designed so that there was inward flow of air through all openings between the manufacturing area and the surrounding environment. This air had to be exhausted from the room in order to achieve directional flow. Before it was discharged to the atmosphere, it was filtered through HEPA filters to remove all traces of penicillin.

Aseptic packaging of pharmaceutical products does not represent the volume of material handled in food processing. However, in food processing the objective is to provide a desired extended shelf life, while in the case of pharmaceuticals it is to prevent the introduction of any particulate material, especially viables that could damage the product and make it unsafe to use.

The regulations established by the Food and Drug Administration (FDA) are referred to as Good Manufacturing Practices (GMPs). There are several GMPs that cover environmental control in the manufacture of different drugs. One of these GMPs[18] is the to use laminar flow devices and so indirectly mandates their use.

In aseptic packaging of drugs, air cleanliness is only one consideration. The size of the bottle or ampoule neck, and the time the bottle or ampoule is opened are equally important. A paper[18] referencing all these considerations used a formula developed by another researcher and found that, assuming neck size was a constant (0.6 cm^2) and open exposure time was also constant (300 s), the effect of airborne cleanliness on the microbiological contamination of open ampoules was as follows:

Airborne Cleanliness	Contamination Rate
ISO 5	0.127
ISO 4	0.026
ISO 3	0.005

Reducing the concentration of contaminant by 90% creates an 80% reduction in the contamination rate.

Other Applications Involving Microorganisms

The growing of mushrooms is another illustration of the three-step process discussed in the section under food processing. In order to improve the quality and yield of mushrooms, those having the desired characteristics are harvested at maturity with the mushroom veil intact. The veil is the underside of the cap. The mushroom surface is decontaminated by dipping it in a disinfectant solution to destroy any bacteria or viruses which may be clinging to the cap. The decontaminated mushroom is cut up with sterile instruments and the spores removed from the gills. The spores are inoculated into prepared test tubes or Petri dishes containing sterile nutrient jelly.

The laboratory work is done in a Class II Biological Safety Cabinet. The sealed containers with the nutrient and spores are stored in racks in incubators where a "pure culture" of mycelium is formed. The pure culture may be used to produce more pure culture or to produce spawn, the seed material used to produce mushrooms.

The production of spawn is similar to that of starter for milk products, except that it takes place in a greater number of steps. The nutrient is primarily rye grain which has been given pretreatment and then autoclaved to destroy any microorganisms which may be present. Pieces of pure culture are added to flasks containing the nutrient and shaken. The mixture is incubated until the mycelium has penetrated all of the nutrient. This mixture of mycelium and nutrient is known as spawn.

Portions of the spawn are taken from the flask and used to inoculate liter-size bottles of sterile pretreated nutrient. When the mycelium has grown, this spawn is used to inoculate gallon-size jars of nutrient and the cycle repeated. Inoculation of flasks and liter-size bottles is usually carried on under laboratory conditions. The inoculation of the gallon-size jars is frequently performed under a HEPA downflow module which bathes the whole operation with clean air and prevents contamination of the seed spawn or the nutrient. When the mycelium has penetrated all of the nutrient in the jars, the spawn is transferred to a bagging operation where the material is transferred from the jar to plastic bags. This operation is also performed under a HEPA downflow module.

In the commercial growing of mushrooms, spawn is added to specially prepared and pasteurized compost in beds of different types depending on the type of growing room used.

Air to growing rooms is usually cleaned using MERV 14 filters. However, some mushroom beds have been destroyed by an airborne virus. As a result, many air supply systems are now filtered with HEPA filters. HEPA filters are also used on the exhaust from the growing rooms to prevent the spread of viruses which may have gotten into a bed.

Most mushroom growing areas are thoroughly cleaned between crops to control harmful viruses, bacteria, and foreign mold. This cleaning may include steam sterilization. Free moisture and high humidity from the condensation of steam can have a damaging effect on certain types of filters. Care should be taken to make sure that filters selected for mushroom growing rooms or caves will survive any cleaning cycle.

General Overview

Reports of different applications, some of which have been listed above, indicate that the following general rules apply:

> Currently HEPA filters are preferred for the removal of airborne viruses. Airborne bacteria can be removed with filters MERV 14 or higher. Airborne mold and fungi can be removed with filters MERV 13 or higher.

Directional flow should be used to prevent the introduction of an unwelcome organism into an area. Airflow should be out of the area to protect the contents and toward the area to prevent the contents of the area from escaping into the surroundings.

Laminar flow clean benches, biological safety cabinets and HEPA filtered downflow work areas provide localized control to protect people, products, and the environment.

Notes to Chapter 10

1. J. Bruce Harstad, Herbert M. Decker, Lee M. Buchanan and Melvin E. Fuller, Penetration of Submicron Ti Bacteriophage Aerosols and Bacterial Aerosols through Commercial Air Filters, 9th Air AEC Air Cleaning Conference, Harvard University School of Public Health, September 13-16, 1966.
2. Henry F. Allen MD, Air Hygiene for Hospitals J.A.M.A., Vol 170 pp. 261-267, May 16, 1959.
3. H. M. Decker, F.A. Giele, H.E. Moorman and C.A. Glick, Removal of Bacteria and Bacteriophage from the Air by Electrostatic Precipitators and Spun Glass Filter Pads, ASHVE Journal Section of HPAC, Oct. 1951, pp. 125-128.
4. Ibid.
5. Frank B. Engley Jr., The Persistent Survival of Microorganisms. I: Airborne Organisms. (Additional information not available).
6. AIA Guidelines for Design and Construction of Hospital and Healthcare Facilities, 2001. The American Institute of Architects, Academy of Architecture for Health, The Facilities Guideline Institute - with assistance from the U. S. Department of Health and Services, 1735 New York Ave. NW, Washington DC 20006.
7. 2007 ASHRAE Applications Handbook Chapter 7 Healthcare Facilities ASHRAE, Atlanta GA 30323 page 7.3.
8. Guidelines for Preventing the Transmission of Mycobacterium Tuberculosis in Healthcare Facilities, CDC Vol 43/No RR-13.
9. OSHA 29 CFR Part 1910 Occupational Exposure to Tuberculosis Page 10.6 forth bullet.
10. Design, construction and performance criteria for Class II biological safety cabinets as well as a list of products that meet these standards have been developed and are available from National Sanitation Foundation, Ann Arbor MI.
11. Biomedical Research Laboratory Facilities, American Institute of Architects, 1735 New York Ave. NW, Washington DC 20006.
12. 3-A Sanitary Standards, Inc., "3-A Accepted Practices for Milk and Milk Products Spray Drying Systems," No. 607-03 (2002).
13. Ibid, No. 607-04 (2002).
14. Ibid, No. 608-01 (2002).
15. The latest version is 3A 608-01 (Feb. 1988) Instantizing System for Dry Milk and Dry Milk Products.
16. 13.21CFR, "Human and Veterinary Drugs, Current Good Manufacturing Practice in Manufacture, Processing, Packaging or Holding," Federal Register 43FR190: 45014-45089 (Sept 29, 1978).
17. William R. Frieben PhD., "The Effect of Cleanroom Design and Manufacturing Systems on the Microbiological Contamination of Aseptically Filled Products," 1985 Proceedings of the IES, pp. 53-56.
18. Guidelines for Preventing the Transmission of Mycobacterium tuberculosis in Healthcare Settings, 2005 Prepared by Paul A. Jensen, PhD, Lauren A. Lambert, MPH, Michael F. Iademarco, MD, Renee Ridzon, MD Division of Tuberculosis Elimination, National Center for HIV, STD, and TB Prevention.
19. U.S. Department of Health and Human Services DHHS Publication CDC 93-8395 US Government Printing Office, Washington, DC 20402.
20. National Institute of Health, "Supplement to Recombinant DNA Guidelines," Laboratory Safety Monograph, January 1979.
21. National Sanitation Foundation, NSF International, PO Box 130140, Ann Arbor, Michigan, 48113-0140
22. Ibid, Section 5.25.2.
23. Foarde, Karin et. al. ASHRAE Research Project 909 2000, Research Triangle Institute, RTI North Carolina.

11

Airborne Molecular Contaminants (AMC)

Introduction

A gas is a fluid that has neither independent shape or volume, but tends to expand indefinitely, filling the space that it occupies. Air is a mixture of gases consisting primarily of nitrogen (78%), oxygen (21%), and argon (1%).

Vapors are gases which are formed by the evaporation or sublimation of materials which are usually liquids or solids, under normal conditions of temperature and pressure. For instance, water exists in the atmosphere as a vapor in varying concentrations, depending on air temperature and other climatic conditions. Low molecular-weight compounds, such as propane and sulfur dioxide, exist as liquids only under conditions of high pressure. At normal atmospheric pressure, they boil into a gaseous state. Higher molecular-weight compounds, such as the components in gasoline, tend to be less volatile and move to the gaseous state at a much slower rate.

Humans and animals detect the presence of certain airborne chemical compounds through the stimulation of the olfactory organs in the nose. Compounds which cause this stimulation are said to have odors. Odors can be both pleasant (perfumes, cooking odors, etc.) and unpleasant (rotten eggs, sewage odors, etc.) depending on the reaction to the stimulus. The lowest concentration at which most people can detect, but not necessarily identify an odor (pleasant or unpleasant) is referred to as the threshold level.

A number of sources list odor threshold levels for many compounds and these levels are usually presented as a concentration range. Range can vary by several orders of magnitude. The odor threshold for hydrogen sulfide (H_2S, rotten egg odor) has been reported as being from 0.00047 up to 4.6 parts per million (ppm).[2] For ammonia (NH_3) this level has been reported as ranging from 1 to 46.8 ppm. The more-odorous compounds typically have lower molecular weights. Gaseous compounds with a molecular weight above 300 are generally considered odorless.[3]

Classification of AMC

There are a variety of ways by which AMC may be classified. None are definitive, but the following classifications have relevance to HVAC systems.

Corrosive Gases

Those gaseous compounds which are likely to cause deterioration or damage to the interior of a building or its contents are considered corrosive. These compounds may also have a detrimental effect on human occupants as well. However, the concentrations at which significant corrosion may occur are usually well below levels expected to be of concern to human health. The most common example of a corrosive gas is sulfur dioxide (SO_2). Its presence can cause significant damage to electrical/electronic devices that could result in the eventual failure of these devices. SO_2[4] is also of particular concern in museums, libraries, and archival storage areas due to the potential for destruction or degradation of irreplaceable artifacts.

Irritant Gases

Gaseous compounds which can be said to cause discomfort and potentially permanent damage to an exposed person may be considered irritating. Many of the irritant gases produce symptoms of pain or discomfort to the eyes, skin, mucous membranes, or respiratory system. Ammonia (NH_3), chlorine (Cl_2), ozone (O_3), and formaldehyde (HCHO) are common chemical irritants.

Odorous Gases

Those gases which affect primarily the olfactory senses are considered odorous and usually carry negative connotations. When asked to describe an "odorous compound," most would describe it in terms of unpleasant odors. However, even odors generally considered as pleasant or unobjectionable when present at normal concentrations may quickly become objectionable at higher concentrations. Recall the last time you entered an elevator with someone wearing a lot of cologne.

Because of the wide variation in the ability of people to perceive odors, there is a correspondingly wide subjectivity in the determination of just what constitutes an objectionable odor. The HVAC design engineer will always have to contend with a finite portion of the general population whose ability to perceive odors is greater than the rest. Odors can be caused by simple inorganic compounds such as hydrogen sulfide and ammonia and organic compounds such as formaldehyde. The most common method of identifying odors has been by type (i.e., cooking odors, bathroom odors, cigarette odors). However, with the increased awareness of the Indoor Air Quality (IAQ) issue, more and more problem odor identifications are being made by chemical composition.

Methods of Control

The three methods of gaseous contaminant control most commonly employed in HVAC systems are source control, ventilation control, and removal control. Source control should always be the first strategy examined. Removing the sources of contaminants prevents them from becoming a problem in the first place. Source control solutions could include eliminating smoking in an area or building, venting the exhaust from a manufacturing process to the outside, or simply moving the source to another area (e.g., moving all laser printers and photocopiers to a separate ventilated room). Many times, however, the source of gases or odors cannot be readily identified and, therefore, cannot be removed. Because of the energy-efficient nature of most of the commercial buildings constructed in the last 15 years, the buildings themselves have become the greatest source of AMC.

When source control is not feasible or practical, such as when the building is contributing a majority of the contaminant load, ventilation control should be the next option. However, as in source control, this may not prove feasible in all cases. Ventilation control involves the introduction of clean dilution air into the affected space. Contaminant levels can thus be reduced below acceptable threshold levels. However, as in source control, this may not prove feasible in all cases. Most ventilation air used for dilution would come from outside the building, and in many of our urban environments today, the outside air does not meet the required criteria for the absence of AMC.[5] If this air is used for ventilation control, one would be substituting one group of contaminants for another and even possibly increasing the total contaminant load in the space. Therefore, the outside air must be treated to remove or significantly reduce the contaminant levels in it before it can be used to reduce the contaminant level in the space.

If it is clear that neither of the two control strategies mentioned above will reduce the level of AMC in the affected space, removal control should be employed. Most interest and research over the last few years regarding the control of AMC has been directed toward the processes involved in removing these contaminants from airstreams. Removal control, as employed as part of an HVAC system, is usually accomplished by physical adsorption and/or chemisorption through the use of various dry, granular, gas-phase (dry-scrubbing) air filtration media. Common filter medias include plain and chemically impregnated activated carbons and alumina, silica gel, and zeolites.

Steady-State Model for Air Quality Control

It is sometimes helpful in identifying the three methods of gaseous contamination control for a single space if a steady-state model is developed to describe this situation. For those who are interested in this advanced application, such a steady-state model is described in Appendix Nine: *"Steady-State Model for the Control of airborne molecular Contaminants."*

Overview of Molecular Air Filtration Principles
In gas-phase, or dry-scrubbing, air filtration as applied in commercial HVAC applications, there are two main processes used to remove airborne AMC. One is a reversible physical process known as adsorption. The other, which involves adsorption and irreversible chemical reaction(s), is termed chemisorption.

Adsorption

By definition, adsorption is the process by which one substance is attracted to and held on the surface of another and is usually described in terms of surface energy per unit area of a solid. Atoms and molecules are held together in a solid by physical forces as strong as valence (chemical) bonds and as weak as van der Waals (physical) attractive forces. Molecules in the interior of a solid are completely surrounded by identical molecules and are subjected to equal attractive forces on all sides. Surface energy is caused by molecules in the surface layer which are subjected to unbalanced external forces. However, these attractive forces do not stop abruptly at the surface. Instead they extend outward from the surface, and when this surface energy is strong enough to overcome the kinetic energy of a passing molecule, that molecule is adsorbed by that solid. The terms adsorbent and adsorbate refer to the solid and its captured molecules. Adsorption can occur wherever a material has sufficient attractive force to overcome the kinetic energy of a gas molecule. The adsorption of cigarette smoke on the interior of an automobile or on a person's clothing is an example of how widespread this process is.

Adsorption is viewed as a surface phenomenon. It is important to understand the significance of this statement. The removal capacity of an adsorbent is directly related to its total surface area. Most think of a surface as referring to only the exterior of an object. However, in a porous solid adsorbent, the surface extends well into the interior of the solid. Therefore, it is important to develop as large an accessible surface area per unit volume as possible. Activated carbons are the most common materials which fulfill this requirement. They are created by permeating a dense, solid, carbonaceous material with a network of internal, submicroscopic macro- and micropores. The number of these pores, the ratio of macro-to micropores, as well as their diameters can be closely controlled to provide a maximum gas removal capacity. See Figure 11.1. Total surface areas for activated carbons range up to 47,500 square yards per ounce (1400 square meters per gram), and sometimes as much as one-half of the individual carbon atoms[6] are exposed and available for adsorption. Many other commercial sorbents such as activated alumina, zeolites, and silica gel also exhibit a relatively high surface area to volume ratio.

Figure 11.1. Activated Carbon Submicroscopic Macro-and Micropores.

The rate at which adsorption occurs is dependent on the time required for the adsorbate to diffuse from the external surface to the interior adsorptive sites. Therefore, this rate is inversely proportional to the size of the adsorbent. The removal capacity, however, depends upon the total available surface area of the adsorbent. Since the internal surface area of the walls of the pores makes up practically all of the surface area of the adsorbent, the capacity is independent of the particle size and, according to one manufacturer, "is, therefore, not enhanced by crushing the adsorbent."[6]

Because of the relatively weak forces holding the adsorbate on the adsorbent, adsorption is (essentially) totally reversible[7] Thus the net rate of adsorption depends on the rate at which adsorbate molecules reach the surface of the adsorbent, the percent of those making contact which are adsorbed, and the rate of desorption. However, many other factors can affect removal of AMC by physical adsorption. Among these are:

- the type of adsorbent
- velocity of the air stream
- the type and concentration of the contaminant(s)
- the temperature and relative humidity of the gas stream.

Although most of the factors listed above can be directly controlled, changes in the temperature and relative humidity of the gas stream may not be easily controlled, especially if outside air is being introduced to a gas-phase air filtration system. Changes in outdoor temperature and humidity are somewhat unpredictable, and can have profound effects on the overall performance of a gas-phase air filtration system through its dry-scrubbing media. Each of these environmental parameters can work to both help and hinder the effectiveness of the media. However, of the two, changes in relative humidity can have the more deleterious effect on the media.

Adsorption occurs more readily at lower temperatures and humidity. Low temperatures reduce the kinetic energy of the gas molecules and the amount of energy required from the system for adsorption. Low humidity reduces the amount of water which is adsorbed by the media. Adsorbed water reduces the capacity of the media for the target gases due to a reduction in the number of available adsorptive sites. The reverse is true for higher temperatures and humidity. Also, at higher temperatures, the rate of gas desorption increases. The adsorption of water (on activated carbon) at various relative humidity levels is shown in Figure 11.2.

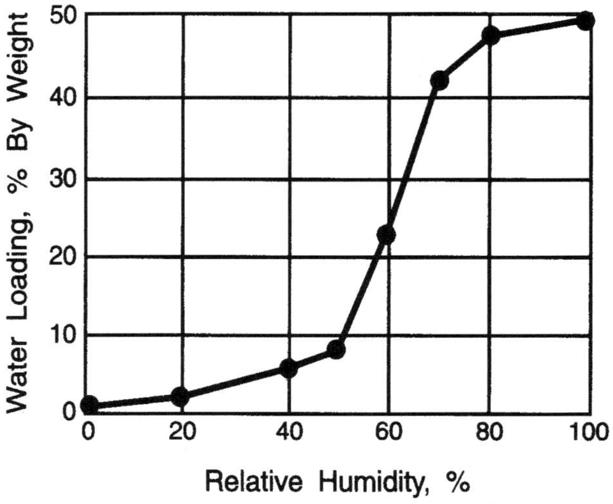

Figure 11.2. Water Adsorption Isotherm for Activated Carbon

As can be seen from Figure 11.2, if the relative humidity of the air stream passing through the gas-phase filtration media is at or below 40%, the water content of the carbon would be less than 5% by weight. At 60% relative humidity, the water content increases to about 25%. At 100% relative humidity, the carbon could have adsorbed up to 50% its weight in water. This moisture equilibration can happen very rapidly depending on the thickness of the media bed and the size of the media. Thin media beds with a small media would be affected more rapidly and severely than thicker beds employing a larger media. Thicker media beds tend to "dampen" changes in humidity and make any adverse effects less severe.

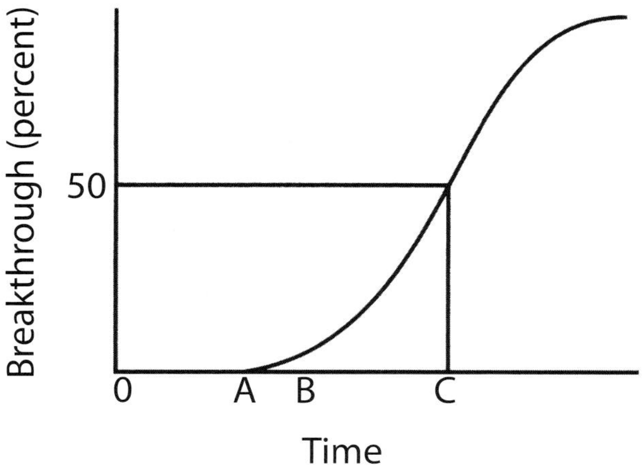

Figure 11.3. Typical Breakthrough Curve for Activated Carbon

Breakthrough Curves. For every combination of adsorbent and adsorbate, it is possible to plot a breakthrough curve which represents the amount, or percentage, of adsorbate which is retained on the adsorbent over time. This may also be described as a removal efficiency curve. Figure 11.3 represents a typical breakthrough curve. Constants which must be identified include adsorbent bed depth, velocity of the air/gas mixture passing through the adsorbent, and the gas concentration in air.

The different labeled points on the curve represent some of the significant events which occur with each adsorbent/adsorbate combination. Point "A" represents the first measurable breakthrough of the adsorbate. The time it takes to go from a 0% breakthrough to point "A" may be very short or fairly long. This time interval is a function of many factors, the most important of which may be the type and amount of adsorbent used. The higher the surface area and larger amount of adsorbent used, the longer this time interval will be.

Point "B" is the first inflection point of the curve. At this point the breakthrough begins to increase at almost exponential rates. If very high removal of the various contaminant gases present is required, point "B" may indicate when the adsorbent media first needs to be replaced. However, if less stringent removal control is acceptable, the breakthrough may be allowed to reach a point further along the curve, such as point "C," before the media would be changed.

Contact Efficiency/Residence Time. One of the main compromises in using sorption (adsorption or chemisorption) for the control of AMC is between the objectives of efficient contaminant control and minimizing the resistance to airflow. This can be achieved by maximizing the contact efficiency of the system. Many manufacturers recommended using granular sorbents in the size range of 4 to 12 mesh (3.33 to 1.17 mm). It must be noted that contact efficiency and removal efficiency are not the same. The contact efficiency is the percentage of the total contaminant molecules which have come into contact with the media. Removal efficiency is that fraction of the contaminant molecules which, once in contact with the media, are removed by either physical or chemical means.[8,9,10]

In a gas-phase filtration system, the contaminant gases must first come in contact with the media before they can be adsorbed. By maximizing the contact efficiency of the system, one can virtually be assured that the maximum removal efficiency of the system is realized. Contact efficiency for a gas-phase air filtration system may be calculated by the following formula:[11]

$$\varepsilon = 1 - (2^{-100n})$$

Where:

ε = Contact efficiency
n = Residence time

Using this formula, a residence time of 0.01 seconds would give a contact efficiency of 50%, 0.02 seconds, 75%, and so on. However, in order to approach a 100% contact efficiency, a residence time of 0.07 seconds would be necessary. It must be understood that maximizing the contact efficiency does not necessarily mean 100% contaminant removal efficiency. All it means is that the system has the opportunity to operate at its own particular maximum efficiency. Residence times used for these calculations are more correctly termed "superficial" residence times. This is because they represent the time it takes air to cross a distance equal to the thickness of the filter without accounting for the resistance of the media through which it travels.

Mass Transfer Zone/Critical Bed Depth. The Mass Transfer Zone (MTZ) of an adsorption system is the area of the media bed where active adsorption is occurring, and is typically defined as a zone bounded by fully saturated adsorbent (0% efficient) on one side and fully available adsorbent (100% efficient) on the other. The MTZ is strongly dependent upon the air velocity, the properties of the subject contaminant, and its concentration. It is important that the packed-bed be thick enough so that a full MTZ can develop. If the MTZ is longer than the bed depth, the maximum efficiency at which that system can operate may be significantly lower than what would be acceptable for IAQ applications.

It has been reported that, for contaminant gas concentrations common to IAQ applications, bed depths of 1 in. and/or residence times of 0.1 seconds are sufficient to contain the MTZ.[11,12] It should be mentioned that this applies to packed-bed gas-phase media systems only.

Removal Efficiency. It was previously stated in this overview section that the rate of adsorption was inversely proportional to the size of the adsorbent media. Therefore, considering media size alone, smaller media would be more efficient because the time required for the adsorbate to diffuse from the external surface to the interior adsorptive sites would be less. A powder would be the most efficient form an adsorbent could take. However, one cannot effectively move air through a packed-bed of powdered adsorbent. Therefore, one must either increase the size of the media to realize an acceptable pressure drop, or find a way to apply the powdered media in the system.

Adsorptive Capacity. Another statement previously made was that the removal capacity of an adsorbent depends on the total available surface area and is independent of particle size. It is independent of particle size because practically all of the available surface area for adsorption consists of the internal surface area of the walls of the pores. Thus, media of different particle sizes could remove the same amount of contaminant on a percent by weight basis.

The adsorptive capacity of a media is also a function of contaminant concentration. It is directly proportional in as much as the higher the contaminant concentration, the more which will be adsorbed—up to the adsorptive capacity of the media. This is due to the increased pressure of the gas forcing adsorption to occur deeper in the media.

Summary: Efficiency and Capacity

The word "efficiency" can be used in several ways when discussing the principle of adsorption. As a result, it is best if it is modified so that the specific function being referenced is identified.

Contact efficiency. The percentage of total contaminant which have come in contact with the adsorbent media. For the contact efficiency to approach 100% the residence time of a packed bed must be .06 to .07.

Residence Time. The amount of time it takes the contaminated air to pass through the filter bed without accounting for the resistance of the media through which it travels.

Note: Maximizing the contact efficiency does not necessarily mean 100% containment removal efficiency. All it means is that the system has the ability to operate at its maximum efficiency.

Adsorption efficiency. A measure of the probability that once a gas molecule has come in contact with an adsorbent particle, it will be retained by that particle. This requires that the gas molecule move from the particle's external surface into interior adsorptive sites. The smaller the adsorbent particle size, the faster the gas molecule can move into an adsorptive site. For this reason, adsorption efficiency is inversely proportional to adsorbent particle size. This statement presumes that all other variables which can effect adsorption efficiency are constant.

Removal capacity. Depends on the total available surface area of the adsorbent. It is a measure of the total amount of gas which the adsorbent will capture and hold. If equal weight samples of two different sizes of the same adsorbent are exposed to an airstream containing an adsorbable gas until the samples can adsorb no more of the gas, the weight increase will be the same for both samples. The smaller size adsorbent particles will become saturated more quickly because their adsorptive efficiency is greater. However, since the total available surface area of the adsorbent is essentially independent of particle size, the larger adsorbent particles will adsorb as much of the gas as the smaller particles, but will not do it as quickly.

Chemisorption

Adsorbent materials do not adsorb all contaminant gases equally. For instance, high molecular-weight (80) and gases with high boiling points (>100F) are preferentially adsorbed over lower molecular-weight and gases and low boiling points when present in the same airstream. Also, if these lower-weight/low boiling point gases were initially adsorbed, they may be desorbed by the introduction of a higher molecular-weight molecule. If the sorbent bed is of sufficient depth, these displaced molecules may again be adsorbed. However, at some point, they would be released back into the airstream. One way to improve the effectiveness of sorbents for these materials is by the use of various chemical impregnants which will react with these "less adsorbable" gases. These impregnates react essentially spontaneously and irreversibly with these gases forming stable chemical compounds which are bound to the media as organic or inorganic salts or released into the air as carbon dioxide, water vapor, or some material more readily adsorbed by other adsorbents. Therefore, it is not uncommon to have a gas-phase air filtration system which uses a combination of unimpregnated and chemically impregnated adsorbent medias.

In contrast to the reversible process of physical adsorption, chemical adsorption, or chemisorption, is the result of chemical reactions on the surface of the adsorbent. Chemisorption is specific and depends on the chemical nature of both the adsorbent and the adsorbate. It is actually a two-stage process. First the adsorbates are physically adsorbed onto the adsorbent. Once adsorbed, they react chemically with the adsorbent. Some oxidation reactions have been shown to occur spontaneously on the surface of the adsorbent. However, a chemical impregnant is usually added to the adsorbent which makes it more or less specific for a contaminant or group of contaminants. For example, activated carbon may be impregnated with sodium or potassium hydroxide (NaOH, KOH) for the removal of acid gases such as sulfur dioxide and hydrogen sulfide. Although some selectivity is apparent in physical adsorption, it usually can be traced to purely physical, rather than chemical properties. In chemisorption, stronger molecular (valence) forces are involved. Many of the same factors which affect the removal of gases by physical adsorption also affect their removal by chemisorption.

The chemical reactions which are occurring during chemisorption are favored, to an extent, by higher temperatures and humidity. Higher temperatures increase the rates of reaction. The extra water enhances the ability of the adsorbed gases to contact the chemical impregnant. Lower temperatures and humidity have the opposite effects. Gases which are adsorbed but not chemically reacted are subject to the same effects of temperature and humidity as for nonimpregnated activated carbon. Desorption of these gases is also possible. The effects of varying humidity on a chemically impregnated gas-phase media is shown in Figure 11.4.

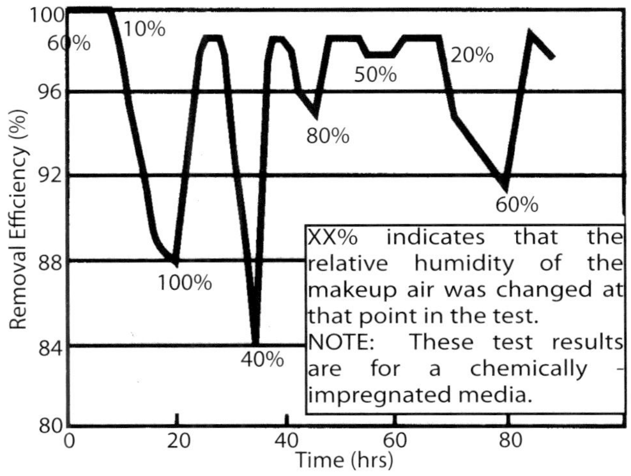

Figure 11.4. Humidity Effects on Gas-Phase Air Filtration Media

Lowering the humidity of air passing through a chemically impregnated media lowers the efficiency. Once the humidity is increased, the efficiency begins to recover. However, if the humidity remains at high levels, the efficiency again begins to decrease due to the adsorption sites now being occupied by water. Once the humidity is again decreased, and the moisture content of the media approaches optimum levels, the efficiency will again increase. Figure 11.4 illustrates the changes in efficiency which could be expected due to the day-night, as well as seasonal fluctuations of relative humidity.

Chemisorption may be employed whenever physical adsorption by itself is inadequate or ineffective against a particular contaminant or group of contaminants. Desorption of target contaminants, once adsorbed and chemically reacted, does not occur.

Catalysis

Chemisorption may be thought of in terms of converting an objectionable or hazardous chemical contaminant to less objectionable or nonhazardous materials, or as the conversion to a material which is more easily adsorbed. Through chemisorption, for example, formaldehyde may be totally oxidized to carbon dioxide and water vapor. The amount of formaldehyde which can be chemically converted is finite, however, based on the amount of chemical impregnant present.

Catalysis may also be thought of as a conversion process. However, the difference between catalysis and chemisorption is that a catalyst is not consumed, or used up, during the conversion. A catalyst is any substance of which a fractional percentage notably affects the rate of a chemical reaction without itself being consumed or undergoing a chemical reaction.[13] Most catalysts accelerate reactions, but a few retard them. Catalysts may be inorganic, organic, or a complex of organic groups and metal halides. They may be either gases, liquids, or solids. In some cases their actions are destructive and undesirable, as in the oxidation of iron to its oxide (rust), which is catalyzed by water vapor. The life of an industrial catalyst varies from 1,000 to 10,000 hours, after which it must be replaced or regenerated.

Since the activity of a solid catalyst is centered in a small fraction of its surface, the number of active sites can be increased in much the same way as the activity of an adsorbent is (e.g., by increasing the porosity and/or the surface area of the material). Catalytic activity can be decreased by substances that "poison" the catalyst by clogging or weakening the catalytic surface. This is why leaded gas cannot be used in cars with a catalytic converter controlling exhaust emissions. The lead poisons the catalyst.

Catalysts are highly specific in their applications; hence their applications in gas-phase air filtration are limited. Also, for a high-contaminant removal efficiency to be realized, several preconditions may have to be met. Among the most common of these is using the catalyst material in a dry airstream or at elevated temperatures.

Molecular Filtration Media

Activated Carbon

The most common material used in HVAC systems for the removal of AMC is activated carbon. A variety of materials such as coal, (petroleum) coke, wood, and (coco)nut shells are used in its preparation. Each raw material imparts its own special characteristics to the activated carbon which, in turn, may make it the preferred product for a specific application. Although many different grades of activated carbon are available, there are a number of different physical properties which can be measured and used for determination of the applicability of various carbons for different uses. Some of these will be described below.

CTC Activity. The CTC, or carbon tetrachloride (CCl_4), activity as measured by a procedure developed by the American Society for Testing and Materials (ASTM), is basically a measure of the pore volume of an activated carbon sample. The original procedure used ASTM Method 3467-88 that required carbon tetrachloride. This has been replaced by ASTM D574295 that measures the butane activity. There is a standard method for converting butane activity to CTC activity. The latter figure is still used even though it is based on butane activity. This test is therefore a means of determining the degree of completion of the activation process, hence a useful means of quality control for gas-phase activated carbons.

ASHRAE SPC 145 is currently working on a proposed standard for testing of media and filters.

Particle/Granule Size. Although it can be demonstrated that, everything else being equal, smaller-sized adsorbents will be more efficient in adsorbing gases, the size of the carbon granules used in any given gas-phase air filtration system must be optimized for that system. The carbon must be small enough to reduce the possibility of by-pass around, or through the media, but large enough not to cause excessive pressure drops through the media. In addition to size itself, size distribution must also be considered. Pressure drops will be the lowest when (roughly) spherical, uniform-sized media is used. A typical gas-phase carbon may be described as being 90% of a 4 6 mesh. This means that 90% of the media will pass through a No. 4 mesh screen and be retained on a No. 6 mesh screen.

Hardness. If two activated carbons have the same CTC activity rating, the hardness of each material may help differentiate between the two samples. If one carbon is softer than the other, it may be subject to excessive dusting during routine handling and use. ASTM test method 3802-79 (1986) describes the test method most commonly used for the determination of the hardness of a carbon sample.[14]

Density. Another physical characteristic of activated carbon which is important in determining the true performance capabilities of a gas-phase system is its packing density. This is usually reported in terms of grams per milliliter (g/ml) or pounds per cubic foot (lb/ft^3) In most comparisons of different carbons, it does not matter which terminology is used, as long as one or the other is used for each carbon to be evaluated. The importance of density can be illustrated in the example below.

Adsorptive Capacity. If it is accepted that the removal capacity of an adsorbent depends on the total available surface area and is independent of particle size, then two carbon medias of different particle sizes, exhibiting the same CTC activity, when tested for total contaminant removal capacity, should be able to remove the same amount of the same contaminant on a percent by weight basis. Adsorptive capacities of activated carbon for hundreds of AMC have been determined and are usually published in the form of adsorption isotherms. These show the amount of contaminant, usually as a percent of the weight of carbon, which is removed at various concentrations, usually expressed in parts per million (ppm) or partial pressures of gas.

Example: Assume one wishes to evaluate two activated carbon samples for use in a gas-phase system. Reported physical characteristics are as follows:

Carbon Sample	"A"	"B"
Density (lbs/ft^3)	28	32
CTC Activity (%)	60	40

If the system for which these two carbons are being considered will hold 4 ft^3 of media, then it could contain 112 pounds of "A"or 128 pounds of "B." This would lead some to believe that "B" would give them a longer period before the media would need to be changed. However, this would not be the case. If the density is multiplied by the CTC activity, one would see that "A" could adsorb 67.2 pounds of CCl_4 whereas "B" could only adsorb 51.2 pounds. It could be assumed that this would equate to a higher overall gas removal capacity. Therefore, even though sample "A" has a lower density, its higher CTC activity would make it the preferred media.

Chemically Impregnated Adsorbent Media

As discussed in the previous section on chemisorption, the effectiveness of adsorbents toward lower molecular-weight contaminants and/or non-polar contaminants can be improved by the addition of various chemical impregnants. Since activated carbon is the most common adsorbent, it follows that many chemically impregnated activated carbons are commercially available. Chemically impregnated gas-phase medias include potassium hydroxide

Figure 11.5. Recommended Removal Media for Contaminants

Contaminant	Activated Carbon	Potassium Permanganate Impregnated Media	Caustic Impregnated Carbon	Phosphoric Acid Impregnated Carbon
Acetic acid	•	•	•	
Acetone	•			
Acrolein	•	•		
Amines				•
Ammonia				•
Benzene	•			
Chlorine	•		•	
Ethyl alcohol	•			
Formaldehyde		•		
Gluteraldehyde	•	•		
Hydrogen cyanide			•	
Hydrogen sulfide		•	•	
Methyl alcohol	•	•		
Mercaptans	•	•		
Methylene chloride	•			
Methyl ethyl ketone	•			
Nitric oxide		•		
Nitrogen dioxide		•		
Ozone	•			
Sulfur dioxide		•	•	
Sulfur trioxide		•	•	
Toluene	•			

See Appendix Seven for Additional Contaminant Capacity Index

impregnated carbons for the control of acid gases such as sulfur dioxide (SO_2), hydrogen sulfide (H_2S), chlorine (Cl_2), and hydrochloric acid (HCl); phosphoric acid- or copper salt-impregnated carbon for the control of ammonia (NH_3); or iodine-impregnated carbon for the control of mercury vapors. However, other impregnated medias are also being used in gas-phase air filtration.

While many different chemicals may be impregnated on activated carbon, one of the more broad-spectrum chemical impregnants in common use, potassium permanganate (KMnO4), cannot. It reacts with the carbon itself, rendering the media ineffective in the process. Therefore, it is typically used as an impregnant on inert substances such as activated alumina and zeolite. Potassium permanganate will react with many common air pollutants including sulfur and nitrogen oxides (i.e., SO_2, NO, and NO_2) and formaldehyde (HCHO). A potassium permanganate-impregnated inert carrier is often used in conjunction with a plain activated carbon to provide a very broad-spectrum gas phase air filtration system.

When evaluating different chemically impregnated medias, the same physical parameters of particle/granule size, hardness, density, and adsorptive capacity used for the evaluation of different activated carbons will apply. Additionally, the type and amount of impregnant, as well as the moisture content of the media, is important. However, as with plain-activated carbon, additional testing simulating actual use conditions may be necessary to obtain a meaningful performance comparison.

Emission Factors	Pounds Per 1000 Gallons of Fuel Automobiles	Diesel Engines
Aldehydes (RCHO)	4	10
Carbon monoxide (CO)	2300	60
Hydrocarbons (C_xH_x)	200	136
Oxides of nitrogen (NO_x)	113	222
Oxides of sulfur (SO_x)	9	40
Organic acids (as acetic)	4	31
Particulates	12	110

Figure 11.6. Emission Factors

Catalysts

Some catalysts are of such a nature that they can exist by themselves in useful forms. In other cases they may have to be incorporated into a carrier. Previous reference has been made to the use of chemically impregnated adsorbents to react with contaminants which were not readily adsorbed. Adsorbents can also be impregnated with chemicals which are catalysts. A good example of this is Whetlerite which was developed during World War II for chemical warfare respirator cartridges and collective chemical protective devices. Whetlerite is a carbon impregnated with salts of chromium, manganese, and silver. The major difference between catalyst-impregnated adsorbents and pure catalysts is that the former may adsorb some of the catalytic reaction by-products, while a pure catalyst is either used alone or in an inert carrier.

Media Selection Considerations

Since many plain and chemically impregnated gas-phase filter medias are commercially available, their applicability for a specific use depends on a variety of factors. The manufacturers of these products have published a wealth of information providing guidelines for the selection of the appropriate gas-phase media or combination of medias. Figure 11.5 is a limited list of contaminants and the media recommended for removal.
How this information may be applied to a specific application is discussed below.

Suppose there is an application requiring the removal of automotive (automobile and diesel) exhaust odors from a building due to these fumes being brought in through

Containment Class	Representative Compound	Threshold Limit Value (TLV, ppm)	Odor Threshold (ppm)
Aldehydes	Formaldehyde	1.0	1.0
	Acrolein	0.1	0.2 - 15.0
Carbon monoxide	Carbon monoxide	50.0	n.a.
Hydrocarbons	Toluene	200.0	2.14 - 15.0
	Cyclohexane	300.0	0.41
	Xylene	100.0	0.47 - 200.0
Oxides of nitrogen	Nitrogen dioxide	5.0	5.0
	Nitric oxide	25.0	0.3 - 1.0
Oxides of sulfur	Sulfur dioxide	5.0	0.47 - 5.0
Organic acids	Acetic acids	10.0	0.2 - 2.4
Others	Hydrogen sulfide	20.0	0.00047 - 4.6
	Ozone	0.1	0.1

Figure 11.7. Threshold Values of Selected Compounds

the HVAC system. As shown in Figure 11.6, there are six main classes of AMC which are routinely described in automotive exhaust.[15] Although carbon monoxide is the predominant contaminant in automobile exhaust and the oxides of nitrogen contribute the most to the overall contaminant load from diesel exhaust, they are not necessarily the components which would be most responsible for odor and/or health complaints from those who may have been exposed to automotive exhaust fumes.

Figure 11.7 shows these six classes (and one other), typical representatives of each class, their respective American Conference of Governmental Industrial Hygienists (ACGIH) threshold limit values (TLVs), and their odor threshold values. OSHA PEL and TWA and STEL see NAFA Guideline on Welding Smokes Odor thresholds are included because odor is the basis for many IAQ complaints. The lower the odor threshold, the lower the concentration at which this material can be detected by humans and the lower the concentration at which IAQ complaints may start. Therefore, just because these contaminants may be present at levels below what would be expected to cause health concerns, does not preclude the fact that they may be present at concentrations high enough to generate IAQ complaints and cause IAQ-related problems.

There have been more than 700 chemical compounds isolated from automotive exhaust, many of which pose a higher potential threat to IAQ and human health than those listed in Figure 11.6. Many of these other compounds have TLVs and/or odor thresholds lower than those listed in Figure 11.7.

Activated carbon does a very good job in removing most hydrocarbons, many aldehydes and organic acids, and nitrogen dioxide due in part to its high surface area to volume ratio. It is not particularly effective against the oxides of sulfur, lower molecular-weight aldehydes and organic acids, nitric oxide, or hydrogen sulfide. Information of this type has been published and available for many years through testing performed by the manufacturers of gas-phase air filtration media and through independent studies.

When tested against various hydrocarbons, caustic-impregnated carbon shows only 70 to 80% (and most often less) of the removal capacity of plain-activated carbon. It is for this reason that a chemically impregnated activated carbon media alone is not generally recommended for the control of automotive exhaust. Rather, it is a gas-phase system employing plain-activated (unimpregnated) carbon and potassium permanganate-impregnated media that offers a good effectiveness against the broadest range of chemical contaminants. Although this example was intended to illustrate how different gas-phase air filtration medias may be applied to a particular problem, there are a number of factors which affect the overall performance of a gas-phase system. Among these are airflow, contaminant types and concentrations, temperature, and relative humidity.

Molecular Air Filtration Equipment Designs

Gas-phase air filters based on adsorption and/or chemisorption are available in a variety of commercial designs[16] usually as packed-bed media filters where the dry, granular gas-phase media is filled in the space between perforated metal or plastic screens. These include units in which a variety of filter bed types and depths are employed. For HVAC systems these include:

- ***Partial Bypass:*** This term typically refers to filters in which the media is loosely packed and the filter itself has numerous voids. Voids are so extensive that one may be able to see through the filter in some cases. Partial bypass filters have low contact and low efficiency and are used primarily for odor control. Since air travels in the path of least resistance, it will preferentially flow through the filter voids. Thus the likelihood of effective contaminant removal is small. Bed depths available include 1, 2, and 4 in. (25, 50, and 100 mm). See Figure 11.8A.
- ***Serpentine:*** These contain thin-bed [(0.375 to 0.5 in.) (9.4 to 13 mm)] convoluted filters and are used in industrial applications with face velocities not to exceed 250 fpm. A typical application is a paint spray booth. See Figure 11.8B.

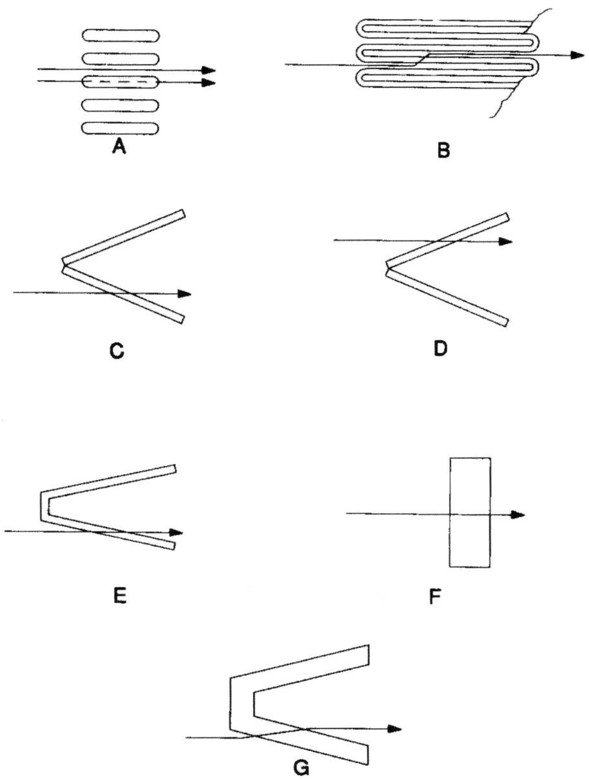

Figure 11.8. Filter Bed Designs Used in Association with HVAC Systems

- ***Refillable trays:*** Used in Side access or front access housing, they are available in depths of .625 " to 1" in the 10 or 12 tray configuration or in 2" thicknesses with a 6 tray arrangement In both cases every 24 x 24 face area is rated for a maximum face velocity of 500 fpm.

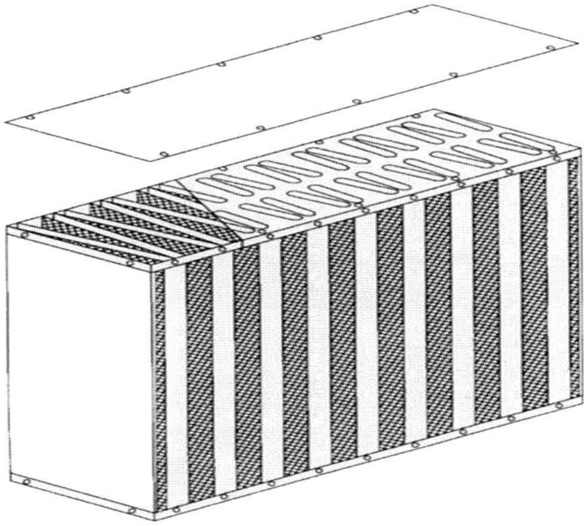

Illustration Courtesy Cameron Great Lakes, Inc.

Figure 11.8C. Standard Refillable Tray Cell

- ***Refillable or disposable*** 18" deep modules with bed depths of 1". These modules are made in a "V" configuration and are suitable for both front and side access housings. Standard 24 x 24 face area requires the use of 4 modules and are rated for a maximum face velocity of 500 fpm. 11.8D

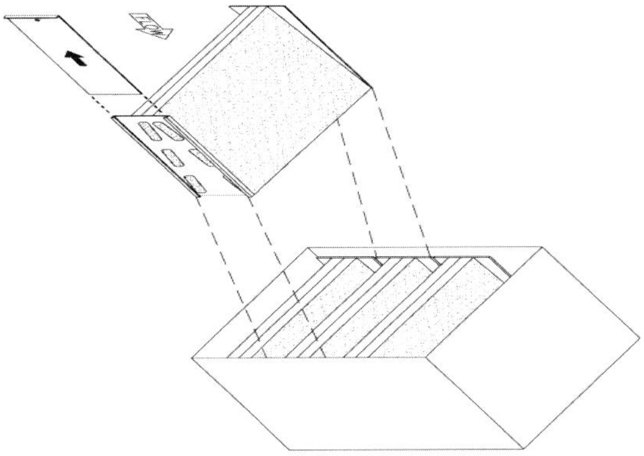

Illustration Courtesy Cameron Great Lakes, Inc.

Figure 11.8D. Refillable Tray Media

- ***Disposable "V" cell Modules:*** Filters are in multiple "V" configuration available in depths up to 12" to attain low bed velocity by using extended surface area techniques. See Figure 11.8E.

- ***Thick-Bed Cells:*** These devices can be as thin as 2 inches and as thick as is required by the contaminant removal specification; some bed depths are 10 feet or more in length found in petrochemical refineries and paper mills. See Figure 11.8F.

Photo Courtesy Cameron Great Lakes, Inc.

Figure 11.8E. Disposable "V" Module

- ***Thick-Bed Depth "V" Module:*** This unit has an average bed depth of 3 in. (75 mm) and is similar to the intermediate depth "V" module. These modules are rated for a maximum face velocity of 250 FPM. See Figure 11.8G.

Residential air cleaners utilize a more limited series of designs. In most instances, thin-bed filters oriented perpendicular to airflow are employed. These filters vary in depth from 0.5 to 2 in. (12.5 to 50 mm). In-duct devices contain thin-bed filters housed in modules similar to those used in HVAC system applications.

Importance of Prefiltration

It is extremely important that gas phase filters be adequately protected from particulate contamination. Gas phase filters should be pre-filtered with particulate filters having a minimum Merv 8. In areas of heavy particulate contamination, filters with Merv 11 or more will be beneficial.

Excessive particulate contamination will adversely affect the life of carbon or any other adsorbent media bed.

Packed-Bed vs. Carbon Impregnated Fiber (CIF) Filters

For many years, packed-bed gas-phase air filters, used in combination with particulate filters, have proven to be very effective for the control of contaminants which contribute to poor IAQ. This two-stage system allows maximization of particulate control and gas (odor) removal. Some felt that this system could be improved upon by combining these two forms of filtration. By bonding a gas-phase media with a particulate filter, it was felt one could offer the same level of contaminant control at a significantly reduced cost. Although these combination filters, commonly referred to as carbon-impregnated fiber (CIF) filters, have proven to be effective in some applications, this strategy has some major shortcomings. These shortcomings are not necessarily in the products themselves, but rather how they are being applied for the control of AMC.[17]

For this discussion, packed-bed filters are considered as those which employ a dry granular, gas-phase (or dry-scrubbing) air filtration media which is bulk-filled into either serpentine, flat panel, or "V-bank" filter assemblies employing (typically) a 1 in. (25 mm) thick media bed. Common filter medias include activated carbon, chemically impregnated activated carbons, zeolites and alumina, and silica gel. These filters are referred to as "packed-bed" due to virtually 100% media particle-to-particle contact. Packed-bed filters allow essentially all of the contaminated air to contact the dry-scrubbing media before exiting the filter.

Photo Courtesy Filtration Group, Inc.

Figure 11.9. Carbon-impregnated Fiber (CIF) Filters

The combination gas/particulate filters use the same types of gas-phase medias as packed-bed filters, but typically have less than 10% of the amount contained in comparable-sized packed-bed systems. These filters employ the gas-phase media either mechanically or physically bound to a fibrous filter substrate in thicknesses ranging from 1/8 in. (3 mm) to over 2 in. (50 mm). Most often this substrate is the same as that used in particulate filters. See Figure 11.9. These combination filters are generically known as (activated) carbon-impregnated medias, carbon-impregnated fibers or foams, or dry-processed carbon composites. However, they all may be generically referred to as CIF filters. While some of the CIF filters are low efficiency others offer a very high initial efficiency. It is important when selecting a CIF filter that you select the proper type for your application.

One of the main features of a packed-bed system is the amount of gas-phase media available for the removal of gases and odors. Typical IAQ applications commonly show media service lives of 1-1/2 to 2 years. This amount of media and the depth of the media bed allow for both high-removal capacities and high-removal efficiencies for the offending gases and odors.

CIF filters are intended to provide the advantages of gas-phase filtration without the associated equipment and replacement media costs and maintenance concerns. Even in retrofit applications these combination filters can be beneficial by using existing hardware and not increasing the demands on the HVAC system with large increases in pressure drops (ΔP) across the filters. Many times, the pressure drops associated with these combination filters are not much higher than the particulate filters they are replacing.

Maintaining a low pressure drop across these CIF filters necessarily means that one cannot incorporate a large amount of gas-phase media into the filter. Therefore, the actual amount of gas-phase media available is significantly less than the amount contained in a typical packed-bed system. This means that the effective service life of these filters would also be significantly reduced.

Service Life of Molecular Air Filtration Media

After a gas-phase air filtration system has been purchased and installed, what one ultimately wants to know is, "What is the expected service life of the filter media?" Service life can be defined many ways, but is most often referred to in terms of removal efficiency (i.e., when the efficiency of the system drops below a minimum value, the gas-phase filter media is replaced). Determination of acceptable minimum efficiency is related directly to the application. What It may be an actual value (e.g., 50%) or can be based on a subjective criteria such as odor. Regardless of how service life is defined, it would be of considerable benefit if a relatively simple expression could be used for this purpose. Two mathematical models have been developed which are being used to predict service life. They are described in Appendix Eight: *"Service Life of Gas-Phase Air Filtration Media."*

Mathematical models can be used for estimations of media service life and to compare competitive products. They are really not much help once the system is installed and operating. As mentioned earlier, the media life in a gas phase filter system depends on many variables. In the absence of proper system maintenance, the first indication that the media may be spent and needs to be replaced might come from building occupants themselves, in the form of complaints pertaining to odors or irritation. Because of the subjective nature of many IAQ problems, this may be the most reliable indicator of media life. However, one would hope that through preventive maintenance, media can be replaced on a timely basis in order to head off IAQ-related complaints.

Some media manufacturers offer media analysis services for the expressed purpose of assisting their customers in determining the proper media change out intervals. A sample of the media is collected from the system and sent to the manufacturer for analysis. The analysis typically involves a determination of the activity remaining in the sample, comparing that to new media, and using the time in service to estimate how much longer the media may be effective. This technique has proven to be particularly effective for those chemically impregnated medias due to the relatively simple analyses involved in determining the amount of unreacted impregnant remaining on the media.

Plain-activated carbons, can be tested for butane activity and this number can be compared to the virgin product, however this is not a completely accurate test as some contaminants may be off gassed during the test. It does however provide some idea of activity remaining and from this serviced life can be estimated.

The most reliable, accurate, and certainly the most expensive way to gauge media life is by the use of gas-monitoring equipment. If critical control of a particular contaminant or group of contaminants is required, this may be the only option.

Disposal of Spent Molecular Media

Once gas-phase media is considered spent, it must be either replaced or regenerated. Spent media used in IAQ applications has traditionally been treated as ordinary commercial waste and disposed of in landfills. Manufacturers can offer assistance on the disposition of their spent media and will supply the end user with a spent carbon profile form which also helps the end user determine if the spent carbon is hazardous, generally the forms for carbon used in HVAC system versus an industrial process are different. Some may offer analytical support for the determination of any hazardous properties of spent media. However, if any question remains as to whether or not spent media may have to be treated as a hazardous waste, one should secure the services of an established and reputable environmental testing laboratory to perform the required analyses. In all cases it is important to observe all local, state, and federal regulations aimed at solid wastes, It is ultimately the responsibility of the users of gas-phase air filtration systems to properly dispose of the spent media generated in their systems.

An option to disposal of spent media is that of regeneration or reactivation. This option typically only applies to plain-activated carbon, some catalysts, zeolites, and silica gels. Chemically impregnated medias are physically and chemically changed through chemisorption and usually cannot be regenerated. Typically, chemically impregnated medias are used only once and then disposed.

Activated carbon can be regenerated through the process of reactivation. In this process, which duplicates the original activation process, the activity of the carbon can be restored to levels comparable to new media. Some physical degradation of the media occurs and any loss of media weight (volume) due to this may be made up with new media. In-place regeneration or reactivation is possible, but not common, in commercial HVAC applications. Typically, the media must be removed from the system, regenerated off site, and replaced.

Some catalysts may also be regenerated if they have become "poisoned." Many times this simply involves drying the media to remove moisture or "washing" the media to remove the chemical reaction by-products. Silica gels and zeolites can also be regenerated in this manner.

Notes to Chapter 11

1. Compilation of Odor and Taste Threshold Values Data, WH. Stahl, Ed., American Society for Testing and Materials, 1973.
2. Handbook on Fundamentals, American Society for Heating, Ventilating, and Air-Conditioning Engineers, Atlanta, 1973.
3. Best Practice guidelines for Libraries, Archives and Museums, NAFA 2005.
4. "National Primary and Secondary Ambient Air Quality Standards," Code of Federal Regulations. Titles 40. Part 50 (4OCFR 50) U.S. Environmental Protection Agency.
5. "Van der Waals equation express. . ." Technical Brochure, Pittsburgh Carbon, Division of Calgon Corporation, 1969
6. J.W Hassler, "Purification with Activated Carbon "Chemical Publishing Co., New York, 1974, pp. 363-367.
7. Turk, H. Mark, and S. Mehiman, "Tracer Gas Nondestructive Testing of Activated Carbon Cells," Mat. Res. and Stds., Vol. 9: pp. 24-26 (1969).
8. Air Pollution. Third Edition. Volume IV - Engineering Control of Air Pollution, A.C. Stem, Ed., Academic Press, New York, NY, 1977 pp. 329-363.
9. R.G. Mathey, T.K. Faison, S. Silberstein, et al., Air Quality Criteria for Storage of Paper-Based Archival Records, NBSIR 832795, Public Buildings Service, General Services Administration, Washington, D.C., 1983, pp. 26-27.
10. J.R. Graham and M.A. Bayati, "The Use of Activated Carbon for the Removal of Trace Organics in the Control of Indoor Air," Proceedings of the Fifth International Conference on Indoor Air Quality and Climate, International Conference on Indoor Air Quality and Climate, Ottawa, 1990, pp. 133-138.

11. R.T. Liu, "Removal of Volatile Organic Compounds in IAQ Concentrations with Short Carbon Bed Depths," Proceedings of the Fifth International Conference on Indoor Air Quality and Climate International Conference on Indoor Air Quality and Climate, Ottawa, 1990, pp. 177-182.

12. G.G. Hawley, The Condensed Chemical Dictionary, 10th ed, Van Nostrand Reinhold Company, Inc., New York, 1981.

13. ASTM Test Method D 3802-79, "Standard Test Method for Ball-Pan Hardness of Activated Carbon," American Society for Testing and Materials, Philadelphia, 1986.

14. Data from Office of Air Programs: "Compilation of Air Pollution Emission Factors," Publication No. AP-42, United States Environmental Protection Agency, Research Triangle Park, 1972.

15. T. Godish, Indoor Air Pollution Control, Lewis Publishers, Inc., Chelsea, Michigan, 1989, pp. 293-298 & 301-303.

16. C.O. Muller, "A Comparison of Packed-Bed Versus Carbon-Impregnated Fiber Filters Used for Gas-Phase Air Filtration: Phase 1," Proceedings of the 86th Annual Meeting & Exhibition, A1219. Air & Waste Management Association, Denver, 1993, pp. 1, 4-5

12

Indoor Air Quality

Introduction

For hundreds of years, the effects of contaminated air on human health has been known. Focused initially on outdoor air, environmental studies showed the potential problems of elevated outdoor air levels of particulate and gas contaminants on health. Early studies on indoor air pollution were focused on the risk to health imposed by smoking indoors.

When the energy crisis occurred in the early 1970's efforts were undertaken to make buildings leak-free so that there would be reduced heating and cooling losses due to the uncontrolled flow of air into and out of a building (infiltration and exfiltration). These losses include not only those due to poorly-sealed windows and doors, but to the building construction itself. Better building shell materials, improved caulking and the use of vapor barriers required for fully-effective use of insulation, helped accomplish the objective of making a building more air-tight. This meant that for many buildings the true ventilation rate (building air changes through all channels) was limited to the ventilation supplied by the HVAC system.

Further reductions in ventilation rates occurred when, as another energy-saving action, the amount of outdoor air was reduced to the then-allowable minimum of 5 cfm (2.36 L/s)(0.00236 m^3 second) per person.

Because air being recirculated in the heating, ventilating and air conditioning system was not diluted with outdoor air as much as in the past, indoor air pollution was allowed to build up. The result was the "Sick Building". Examination of the sick building resulted in a new field of study, "Indoor Air Quality" (IAQ).

Much has been written on the subject of "Indoor Air Quality." It is not possible to discuss all the details involved in this topic. Instead, the intent of this chapter is first to present an overview and then to discuss where air filtration, both particulate and molecular can be used to solve IAQ problems.

The occupant of a space is the principle concern of indoor air quality. Air quality is an important consideration for museums, art galleries, assembly areas and even office buildings whose decor and contents require clean air.

Sick Building Syndrome

Most people are aware of the effects of Sick Building Syndrome (SBS) - lethargy, frequent headaches, eye irritation and fatigue- all of which lessen or disappear after people leave the building. A person may attend a meeting in a room, leave it and later return to be overpowered by the "staleness". This condition is usually caused by a buildup of carbon dioxide, cigarette smoke, body odors and other more subtle sources.

Building Related Illness

Building Related Illness (BRI) on the other hand, is a diagnosable illness whose symptoms can be identified and can be directly attributed to environmental agents in the air. Legionnaire's disease, Pontiac fever and humidifier fever have all been reported by medical authorities.

Identifying Pollutants

In many cases the identification of indoor air pollutants is not easy. However, one can define the basic sources: the building itself, including mechanical equipment; the air surrounding the building; the contents of the building; the operations in the building; and the occupants. To catalog most of the pollutants arising from these different sources would take several pages and might still be incomplete. However, some of the more important sources can be listed.

Buildings

The foremost source of contamination coming from a building itself is frequently the finish applied to the walls and other interior surfaces. Chemicals being used in glues in carpets and furniture, solvents and other agents in paints and wood finishes and mastics used during construction all contribute to indoor air pollution in buildings. Old buildings may have been coated with lead-based paints which, if not properly sealed can chip. The chips can be powdered to form a high-lead dust which can be a hazard. Formaldehyde is a volatile organic compound (VOC) that finds its way into many sealants, fire retardants, and foam insulation. It can off-gas and build up a concentration in inadequately ventilated areas. Other VOC's have been identified in cleaning compounds and room deodorizers.

Building Mechanical Systems

The HVAC system has the job of heating and cooling, adding or removing moisture, and transporting and cleaning the air in a building. Since most newer buildings no longer have windows that open, the HVAC system is the only way to bring air into a building. Systems that are not properly maintained can become sources of air contamination. Low efficiency filters, bypass around filters and other leaks in the filter plenum can allow a build-up of dust on coils and other surfaces. If these surfaces become wet they can be a fertile breeding-ground for mold and mildew. Inadequately designed condensate drain pans can allow a build-up of water, encouraging the growth of bacteria and mold. The moldy odor can spread through an entire building causing SBS.

Equally as objectionable has been the residue that exists when moisture is removed and moldy objects are allowed to dry. This residue can become airborne in the HVAC system and can cause allergic reactions in varying degrees to some of the people who are exposed to this dust.

Air Surrounding Buildings

Outside air is sometimes mislabeled as "fresh air" when in reality it may be contaminated with smoke, high levels of ozone, oxides of nitrogen and carbon monoxide from vehicle exhaust. It may contain sulfur dioxide from boiler stacks where sulfur-bearing fuels are burned. Fine dust from various industrial processes and gases and vapors from chemical plants may also be present.

The location of the outside air inlet of an HVAC system requires serious consideration. It is very important that it not be located directly downwind of any contaminant-generating source. If this precaution is not taken, more air cleaning will be required before this air can be used.

ANSI/ASHRAE Standard 62.1-2004

ASHRAE (the American Society of Heating, Refrigerating and Air-conditioning Engineers) has approved the Standard, "Ventilation for Acceptable Air Quality" (ANSI/ASHRAE Standard 62.1-2004).[1]

The defined purpose of this Standard is "to specify minimum ventilation rates and indoor air quality that will be acceptable to human occupants and are intended to minimize the potential for adverse health effects." It permits two procedures to obtain acceptable indoor air quality: 1) By providing air of the proper quality and quantity to the space, the "Ventilation Rate Procedure," or 2) by achieving acceptable air quality within the space with respect to specifiable contaminants, the "Indoor Air Quality Procedure."

Air Classes and Redesignation

Standard 62.1 – 2004, has several designated air and recirculation into 4 classes that allow for it to be transferred or recirculated:

Class 1: Air with low contaminant concentration, low sensory-irritation intensity, and inoffensive odor. Class 1 air may be recirculated or transferred to any space.

Class 2: Air with moderate contaminant concentration, mild sensory irritation intensity, or mildly offensive odors. Class 2 air also includes air that is not necessarily harmful or objectionable but that is inappropriate for transfer or recirculation to spaces used for different purposes.

Class 2 air may be recirculated within the space of origin. Class 2 air may be transferred or recirculated to other Class 2 or Class 3 spaces utilized for the same or similar purpose or task and involving the same or similar pollutant sources. Class 2 air may be recirculated or transferred to Class 4 space. Class 2 air shall not be recirculated or transferred to Class 1 space.

Class 3: Air with significant contaminant concentrations, significant sensory-irritation intensity, or offensive odor. Class 3 air may be recirculated with the space of origin. Class 3 air shall not be recirculated or transferred to any other space.

Class 4: Air with highly objectionable fumes or gases or with potentially dangerous particles, bioaerosols, or gases, at concentrations high enough to be considered harmful. Class 4 air shall not be recirculated or transferred to any space nor recirculated within the space of origin.

Addendum G to the Standard does not allow for any recirculation by air handling systems or natural ventilation from an Environmental Tobacco Smoke (ETS) area to and ETS-free area.[2]

Air (return, transfer, or exhaust air) leaving each space or location shall be designed at an expected air-quality classification not less than that shown in the following table:

Description	Air Class
Spaces ancillary to Class 2 Spaces	2
Break rooms	1
Coffee stations	1
Private toilet/bath	2
Employee locker rooms	2
Laundry rooms, central	2
Laundry rooms within dwelling units	1
Soiled laundry storage	3
Janitors closet, trash room	3
General chemical/biological laboratories	3
University/college laboratories	2
Paint spray booths	4

Air Cleaning

Standard 62.1 allows for a reclassification of air:

> "If air leaving a space or location passes through an air-cleaning system, the cleaned air may be reclassified to a cleaner classification, using the subjective criteria noted above, with the approval of the authority having jurisdiction."

Standard 62.1 also designates MERV 6 filters for those buildings located where the National Ambient Air Quality Standards (NAAQS) for Particulate Matter (PM)10 is exceeded. Buildings located where ozone levels exceeding the one-hour average concentration of 1.160 ppm, (313 $\mu g/m^3$) "air-cleaning devices for ozone shall be provided ...shall have a volumetric ozone removal efficiency of 40% when installed, operated and maintained in accordance with manufacturer recommendations and shall be approved by the authority having jurisdiction."

Standard 62.1 has many more recommendations to guide engineers and users to better indoor air quality.

The Ventilation Rate Procedure

The Standard states: "This is a prescriptive procedure in which outdoor air intake rates are determined based on space type /application, occupancy level, and floor area. Note: The Ventilation Rate procedure minimum rates are based on contaminant sources and source strengths that are typical for the listed spaces.

The Indoor Air Quality Procedure

The alternate Indoor Air Quality Procedure is a design procedure in which outdoor air intake rates and other system design parameters are based on an analysis of contaminant sources, contaminant concentration targets, and perceived acceptability targets. The IAQ Procedure allows credit to be taken for controls that remove contaminants (for example air cleaning devices) or for other design techniques (for example, selection of materials with lower source strengths) that can be reliably demonstrated to result in indoor contaminant concentrations equal to or lower than those achieved using the Ventilation Rate Procedure.

The use of the Indoor Air Quality Procedure is especially desirable because of the economic benefits which may be involved in reducing the amount of outside air used in an air conditioning system.

One published study[4] reported on an installation at a Veterans Administration Medical Center. The air conditioning system serving one wing had a capacity of 26,000 cfm (12,300 L/s), all outside air. It was modified to include appropriate air cleaning equipment. After the air cleaning equipment was added to the system the amount of outside air was reduced to 4,000 cfm (1890 L/s). The balance, 22,000 cfm (10,400 L/s) was recirculated through the air cleaning

Figure 12.1. Selected Ventilation Rates in Breathing Zone

Application		Occupancy (People/1000 ft^2)	Cfm/person
Food and Beverage Service	Dining Rooms	70	10
	Cafeteria, fast food	100	9
	Bars, cocktail lounges	100	9
Offices	Office space	5	17
	Reception areas	30	7
	Conference rooms	50	6
Retail Stores, Sales Floors, Showroom Floors	Sales	15	16
	Mall Common Area	40	9
	Smoking lounge	70	60
Sports and Amusement	Spectator areas	150	8
	Game rooms	20	17
	Playing floors	30	
	Ballrooms and discos	100	21
Education	Classroom (age 9 plus)	35	13
	Music/Theater/Dance	35	12
	Bedrooms/Living rooms	10	11
	Lobbies	30	10
	Conference rooms/meeting	50	6
	Assembly rooms	120	6

Information Courtesy of ANSI/ASHRAE Standard 62.2 - 2004

equipment. The net annual operating cost savings of this system were calculated to be over $57,000 even when the added power cost due to the resistance of the particulate and gas-phase filter and the annual cost of replacing gas-phase purification media were included.

A second published study reported on an installation at a Domed Stadium.[5] In the section titled ANSI/ASHRAE Standard 62.1-2004 conformance it states the following:

> "The Domes air cleaning systems have three stages: a prefilter (cartridge type), an electronic section (sic, electrical excitation system), and a final cartridge filter having an ASHRAE efficiency of MERV 13. Using air cleaning devices had two benefits. By requiring less outdoor air for ventilation, it reduced the chiller plant load 2350 tons and the heating load by 40 million BTU, which resulted in an annual savings of $800,000 in operating costs. It also states that this combination eliminates odors and particulate matter from the air. Even the atriums designated for smoking and food outlets have no odor of cooking or smoke."

Air Filtration and Treatment for Outside Air

As noted previously, air filtration or other forms of treatment are required: (1) when outdoor air contaminants must be reduced to an acceptable level before being used in an HVAC system or, (2) when indoor air must be filtered or otherwise treated before it can be recirculated back to the occupied area.

If outdoor air is to be used, the concentrations of certain contaminants cannot be exceeded. These contaminants are:

- sulfur dioxide
- total particulates
- carbon monoxide
- nitrogen dioxide
- oxidants (ozone)
- lead

In addition, if outdoor air is thought to contain any contaminants other than those listed above, guidance on acceptable concentration levels can be acquired from reference to the appendix of the Standard which may have this information. The Standard recognizes that there is a difference between long-term and short-term exposures and allows for higher concentrations of the latter if they do not exceed certain time intervals. For instance the long term allowable concentration of sulfur dioxide is 0.03 ppm but can be as high as 0.14 ppm for periods which do not exceed 24 hours.[6]

Manufacturers of the different types of equipment and NAFA Certified Air Filter Speciallists should be consulted for their recommendations as to the specific type of equipment required to reduce airborne contaminants in outdoor air to the level where it can be used in an HVAC system.

Air Filtration and Treatment for Recirculated Air

The selection of air cleaners to reduce the concentration of contaminants within an occupied space again depends on the identification and measurement of the concentration of the contaminant. There are many recognized contaminants, and data about them appears in the appendix of the ANSI/ASHRAE Standard 62.1. There are other materials which may not be recognized as contaminants, or which, while recognized, may not be easily measured.

Building Contents

Formaldehyde is used extensively in synthetic fiber materials which make up wall coverings and furniture fabrics. Many of the plastic leather-like furniture coverings have been made with chemicals (plasticizers) to keep the material soft and to prevent it from cracking. These plasticizers will off-gas from the materials and become a component of indoor air.

Studies of "Sick Buildings" have found that some moldy odors which occupants find objectionable have come from wet carpeting. The moistened starches and other materials in the floor covering are fertile grounds for the growth of mold and mildew. As in the case of mechanical systems, when these fungi dry, the residue can be easily made airborne by foot traffic on the carpeting. This residue can cause allergic reactions.

Other sources can be walls, floors and ceilings that have gotten wet, drawing moisture from underneath the building. Leaking roofs or moisture coming through the walls because of siding problems can also be sources.

All these have caused mold and mildew allergies to become more prevalent in recent years.

Building Operations

Frequently the operations performed within a building are major sources of airborne pollutants. Photocopiers may generate ozone during their operation. Dust from these machines and laser printers can also get into the air. Ammonia used in the developer of blueprint machines can escape into the surrounding areas if the machine is not properly vented. In addition there are many other types of office equipment and materials which can contribute to indoor air pollution.

One group of significant contributors to indoor air pollution is the materials used to maintain building interiors. Cleaners frequently contain solvents which evaporate into the building atmosphere. Many cleaners themselves contribute odors which can build up inside a building.

Building Occupants

The people who occupy a building are also the major contributors to indoor air pollution. People are, in effect, furnaces burning (metabolizing) food to provide the energy needed for all bodily functions. Metabolic activity produces a variety of by-products: carbon dioxide, water, sweat, glandular secretions and organic dust particles from hair, skin, mucus and dead cells.[7] It has been found that in crowded rooms the concentration of body odors will be maintained at a satisfactory level if the concentration of carbon dioxide is kept below 700 ppm above outdoor air levels. The carbon dioxide level is not important of itself but is as an indicator of contaminants from occupants.[8]

Pollution Control

Control of indoor air pollution is elusive. Simplistically stated, one has the option of 1) removing the contaminants at their source, (2) removing them during the "air conditioning" process or 3) reducing them to an acceptable level by dilution with "clean air," namely air which does not contain contaminants.

Source Removal

Capture of an airborne contaminant at its source is the most economical method of control. Once it has been captured, a contaminant may be discharged outdoors by an exhaust system. If this is not allowed because of the nature of the contaminant, it may be captured within the equipment in which it is generated or may be removed by an exhaust system into which air purification equipment has been incorporated. An example of capture within equipment is an air filter built into a photocopying machine to collect toner dust. Blueprint machines can incorporate exhaust systems to discharge the ammonia from the developer out of doors. When this is not possible, a filter canister using a specially-treated form of activated carbon can be attached to the discharge of the fan of the blueprint machine. It will remove the ammonia and allow the exhaust to be discharged back into the occupied space.

One useful approach is to isolate all contaminant-producing operations in one room, where dedicated ventilation equipment will control the contaminants generated in the room and at the same time prevent their dispersal into the rest of the building.

Removal by the HVAC System

The removal of airborne contaminants during the "air conditioning process" is an economical rather than a technical problem. If necessary, HEPA filters can be used to remove the smallest undesirable airborne particulate materials. Chemical adsorbers such as activated charcoal will remove gases to varying degrees depending on the nature of the gas to be removed, see Chapter 11: "*Airborne Molecular Contaminants (AMC)*." Increasing carbon bed depth will improve efficiency. Sometimes activated carbon impregnated with certain chemicals can also improve the overall removal of certain contaminants. In addition to carbon there are other materials such as activated alumina impregnated with potassium permanganate which will remove some undesirable gases. One need only remember that nuclear submarines, because of very efficient air treatment equipment, stay submerged for long periods without the use of any outside air. The same technology is available to any building owner, but it is unlikely that it is affordable.

Dilution Ventilation

The simplicity of the idea makes dilution the ideal way to keep the concentration of contaminants below a specified level. It only requires that air free of the contaminant of concern be introduced into the space until the concentration of the contaminant is diluted below a specified level based on space type, application, occupancy level and floor area. However, if air is to be diluted, some of it must be removed from the space to allow the dilution air to enter. This dilution air must be properly conditioned and properly introduced into the space. The expense of doing this can, in many instances, outweigh the simple answer that dilution seems to offer.

Office Buildings

The well-designed office building provides an attractive environment in which to work. Attractive interior decor contributes to this environment. Although the ban on cigarette smoking in most all offices has eliminated odor, potential health effects of side stream smoke to non-smokers and the buildup of unsightly film on walls and other surfaces, contaminants from outside air and from fine dust and from vapors generated within a building must be controlled. Pedestrian traffic can generate a fine dust from the abrasion of carpeting or other floor surfaces. Off-gassing of upholstery material has already been mentioned.

Museums

Very fine dusts and aerosols can adhere to paintings and other artworks, darkening them over time and eventually requiring careful cleaning by a specialist. Oxides of nitrogen and sulfur dioxide chemically attack the fibers in tapestries and the canvas fabric of paintings. Removal of particulate aerosols usually requires that filters of MERV 14 efficiency or greater be used on air supply systems. The type and concentration of the gaseous contaminant determines the type of molecular filtration system which should be used. For more information see the NAFA Libraries, Archives and Museums Guideline available at www.nafahq.org.

Libraries

While many libraries contain various works of art, their principal purpose is to preserve and protect books. Papers of all types are damaged by acid gases such as oxides of nitrogen and sulfur dioxide. The latter is of special concern and the recommended maximum allowable concentration for sulfur dioxide in a library environment is actually less than that listed for the occupants of such a building. Gas-phase purifiers containing a blend of potassium permanganate impregnated media and activated carbon are recommended for use.

Healthcare Facilities

Today's hospital creates many new IAQ concerns in addition to the same concerns that are involved with other buildings. Isolation rooms must be properly designed and filtered to protect the patient (room under positive pressure) and other building occupants outside the room (room under negative pressure).
New designs for surgical suites allow for proper air flow patterns and (HEPA) filtration modules at the ceiling outlet.

Design and filtration of labs protect occupants and the processes, using exhaust hoods and clean benches depending on the application. In most states there are new higher requirements for filter efficiencies for other areas of hospitals based on standards, guidelines and regulation requirements from varying agencies such as the Center for Disease Control and Prevention (CDC) the American Institute of Architects (AIA), the Department of Health and Human Services (HHS) (see Chapter 10: *"Airborne Microorganisms"*).

Schools

Many of the same issues that have been discussed in previous paragraphs come to play in school buildings many times over because of the increased population in these facilities and because of the concern over children and staff health in addition to litigation-driven issues.

Air filtration upstream of condensate-producing devices shall have an efficiency of at least MERV 6 per ANSI/ASHRAE Standard 62.1-2004 Section 5.9, Particulate Matter Removal.9 The Standard additionally states in section 8.4.1.1, "all filters in air cleaning devices shall be replaced or maintained as specified by the Operations and Maintenance Manual."

Smoking Areas

ANSI/ASHRAE Standard 62.1-2004 Minimum Ventilation Rate Table deleted smoking lounges from the list of occupancy categories. New Federal, State and local regulations regarding restriction of smoking in public and private buildings, has created new language on smoking areas. The Standard states that, "ventilation rate requirements cannot be determined until cognizant authorities determine the concentration of smoke that achieves and acceptable level of risk." This wording came about as a result of the intense pressure of the medical community who feels that no level of cigarette smoke in the air can be deemed "safe." The Standard does state that more ventilation is needed for smoking areas and that air from smoking areas cannot be recirculated back to non-smoking areas.

There are no known filtration devices that completely remove all environmental tobacco smoke (ETS) from the air. Therefore, NAFA recommends that building owners and managers prohibit smoking within occupied spaces and within a distance of 15 feet from building air intakes and opening that might retrain smoke or odors into a building.

ANSI/ASHRAE Standard 62.2 – 2004

New to ASHRAE was the writing of a standard for the residential market. The belief was that leakage and operable windows provided residences with enough ventilation for dilution of contaminants. Studies by EPA and others found this assumption incorrect.

ASHRAE wrote and has approved the Standard, "Ventilation And Acceptable Air Quality In Low-Rise Residential Buildings" (ANSI/ASHRAE Standard 62.1-2004). This Standard covers single-family houses and multifamily structures of three stories or fewer above grade including manufactured and modular houses.

Minimum Filtration

Standard 62.2 sets a minimum air filtration efficiency of MERV 6 or better when tested according to ANSI/ASHRAE Standard 52.2-1999. It further states that, "the system shall be designed such that all recirculated and mechanically supplied outdoor air is filtered before passing through the thermal conditioning components." One of the criteria in the Standard is a pressure drop qualification "...operate at a clean pressure drop no greater than 0.1 in. w.g. (25Pa) unless the equipment is designed or selected to accommodate any additional pressure drop imposed by the filter selection."[10]

Summary

ANSI/ASHRAE Standard 62.1-2004 states that acceptable indoor air quality is, "air in which there are no known contaminants at harmful concentrations as determined by cognizant authorities and with which a substantial majority (80% or more) of the people exposed do not express dissatisfaction."[1]

IAQ issues are an important facet in each one of our lives. Research shows cleaner air provides an indoor environment that is conducive to higher productivity levels, decreased absenteeism and improved health. Air filtration at higher efficiencies in an HVAC system is a key ingredient in assuring cleaner and healthier indoor air quality.

Notes to Chapter 12

1. "Ventilation for Acceptable Indoor Air Quality", ANSI/ASHRAE Standard 62.1-2004, the American Society of Heating, Refrigerating and Air Conditioning Engineers Inc., 1791 Tullie Circle NE, Atlanta GA 30329.
2. Ibid Paragraph 6.1
3. "A Practical Approach to IAQ Investigations", National Coalition on Indoor Air Quality (NCIAQ) Seminar, p.46.
4. C.R. Kennedy and A. Distefano, "Improving Building IAQ Reduces HVAC Energy Costs", H/P/A/C/ pp 75-79 (Nov. 1991)
5. "Innovative HVAC Design Halves Domed Stadiums Cooling Costs." In Heating Piping Air Conditioning, by Richard D. Yelvington, PE, from Heery International, entitled
6. National Primary Ambient-Air Quality Standards For Outdoor Air. U.S. Environmental Protection Agency.
7. Robert T. Hughes and Dennis M. O'Brien, "Evaluation of Building Ventilation Systems", JAIHA, 47(4) 207-213 (1986)
8. ANSI/ASHRAE Standard 62.1 "Minimum Ventilation Rates In Breathing Zone."
9. ANSI/ASHRAE Standard 62.1-2004 Section 5.9
10. ANSI/ASHRAE Standard 62.2-2004 "Ventilation and Acceptable Indoor Air Quality in Low-Rise Residential Buildings."

13

Owning and Operating Costs

Note: Energy Cost Consumption used assumes a system using a variable speed drive.

Introduction

For any air filtration system, filtration efficiency comes first. Cleaning the air is the primary function of any air filtration system and the level of air cleanliness should be the first decision that is made regarding a new filter installation, Once a decision has been made regarding the degree of air cleanliness that is required, the next decision is to select a filtration system that will provide the least expensive owning and operating costs – i.e. the system with the highest sustainability for the system, equipment, facility and fixtures.

When two or more systems can provide the same level of air cleanliness but are different in one or more respects, an owning and operating cost comparison can guide the end user in making a choice as to which system to use. Such an analysis will consider the following factors:

- Initial cost – including shipping
- Installation cost – labor to remove/install
- Operating cost – energy cost
- Filter life – based on pressure drop
- Disposal cost

New System/Replacement System Initial and Installation Costs

The first cost for any new or retrofit air filtration system is in the purchase and installation of the equipment. This includes:

- equipment cost (including inbound freight)
- equipment installation charges
- electrical services (automatic roll filters and electronic air cleaners) and plumbing (wash-in-place electronic air cleaners)
- initial filter cost
- filter installation costs
- miscellaneous costs such as gasketing, holding clips and caulking

Operating Costs

Operating costs of an air filtration system are the costs associated with keeping the system operational. Ongoing operating costs of an air filtration system include these three main factors:

- energy costs
- maintenance costs
- disposal costs

The energy consumed by the filter is largest portion of the ongoing operating costs, representing approximately 80% of the cost associated with a air filter system as illustrated in Figure 13.1.[1] Investment and maintenance costs, which account for replacement filters and associated installation charges, represents approximately 18.5% of the total operating cost of a filter system. Disposal costs represent approximately 0.5% of the cost of operating an air filtration system.

The percentages in Figure 13.1 can be verified by completing the energy calculations for various filter types. A summary of this information using typical filter pressure drops is included in Figure 13.2.

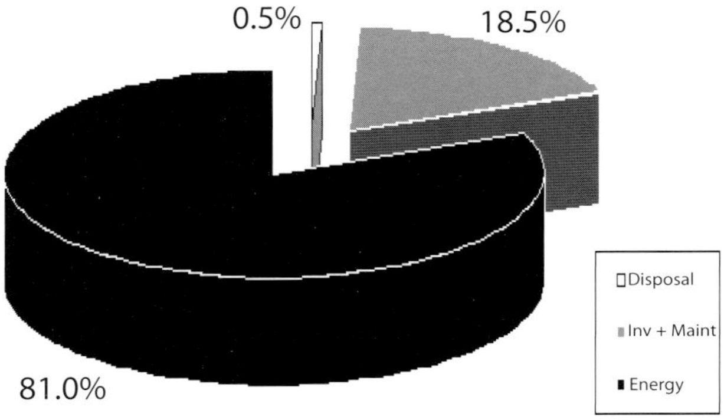

Figure 13.1.

Energy Costs

Energy has become an important factor in the selection of an air filtration system, especially as energy costs rise. The energy that is consumed by an air filter system is the energy needed to overcome the resistance to air flow which is imparted by the use of an air filter. The use of an air filter does impact the heating and cooling capability of an HVAC system with respect to the HVAC coils and system cleanliness, but the impact on these costs is much harder to quantify. This section focuses on the energy that is consumed by the fan to overcome the resistance of the filter while supplying the specified amount of clean air to a building.

The energy used to overcome the resistance of an air filter bank is provided by the blower that is part of the HVAC system. The blower, in turn, gets its energy from a motor. It is rare that this motor is not electric, so the energy consumption associated with air filtration is generally referred to as electrical energy consumption.

	Initial Price	Energy Cost	Initial Cost %	Energy Cost %
Pleated Filter	$4	$34	11%	89%
Bag Filter	$35	$196	15%	85%
Box Filter	$70	$229	23%	77%

A1 energy cost calculations assume: 24/7/365 operation, $0.08kWh energy cost, avg ΔP is line average between initial ΔP and final ΔP, fan, motor, and drive efficiency total 58%, system airflow is 2000 cfm/filter.

Figure 13.2.

The formula for fan energy consumption is:

Energy Consumption =

$$\frac{Q \times \Delta P \times t}{n \times 1000}$$

where:

Energy Consumption in kilowatt hours (kWh)

Q = airflow in cubic meters/second (m3/s)

ΔP = the average pressure drop across the filter in Pascals (Pa)

T = the time the fan is in operation in hours (hrs)

n = the product of the fan, motor, and drive efficiency in %

The formula is very straightforward when using metric units.

Airflow: 1 m3/sec = 2120 cfm

Pressure Drop: 1" WG = 249.09 Pa

The average pressure drop of a filter in use is often difficult to predict. If gauges are used to monitor the resistance of the filters, the average pressure drop can be calculated from historical data. The graphic comparison of the energy required by two different air filtration systems can

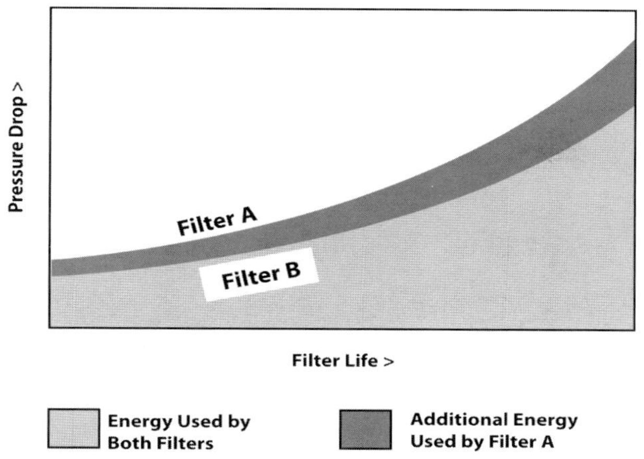

Figure 13.3.

be made by drawing life curves (filter operating hours vs. filter resistance) for each of the filters and comparing the areas underneath the life curves (Figure 13.3). The filter life curve with the larger area under it uses more energy. If there were mathematical formulas to define precisely the curve of pressure drop vs. operating time, the formulas could be integrated and energy usage easily calculated. Unfortunately there are no such equations and so any calculations must be done graphically.

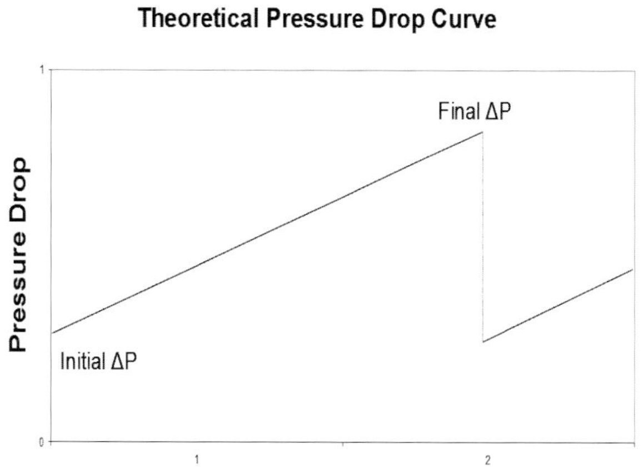

Figure 13.4.

If no historical pressure drop data exists, the theoretical straight line pressure drop curve can be assumed. This is illustrated in Figure 13.4. In this case, the average pressure drop is taken as the average between the initial pressure drop and the final recommended pressure drop as calculated with the equation:

Average pressure drop =
(Final Pressure Drop + Initial Pressure Drop ÷ 2)

Motors and fans are not 100% efficient, and thus, the loss of power must be accounted for. The fan, motor, and drive efficiency is important because it factors in the power loss that occurs in the system prior to the energy being transferred to the air. By factoring in an efficiency number for the fan motor and drive, we are able to account for the power loss in the transfer of energy through the v-belt drive between the motor and the fan. One commercial publication[3] suggests that:

Fan Efficiency = 68%

Drive Efficiency = 99%

Motor Efficiency = 86%

Electrical energy is sold in the form of kilowatt hours (kWh). For an air filtration system, the energy cost associated with the energy used is calculated by multiplying the energy consumption by the energy cost per kWh as indicated below:

Total Energy Cost ($) =
Energy usage (kWh) x Energy Cost ($/kWh)

Maintenance and Disposal Costs

Maintenance costs include all the expenses required to keep an air filtration system operating at an acceptable level after the initial installation. For replaceable unitary filter systems, maintenance costs include new filters, the labor to bring replacements to the filter bank, the removal of the spent filters, cleaning and checking of holding frames, gaskets and fasteners, the installation of the replacement filters, the transfer of the spent filters to the disposal area, and the disposal cost.

Other air filtration systems may have different maintenance costs. Electronic air cleaners usually have either removable or wash-in-place elements. Removable elements must be taken to an area where they can be washed and dried and then reinstalled before the filter system can be put on-line.

Wash-in-place electronic air cleaners are usually programmed to go through the whole wash-rinse-dry program automatically. While this work can be done unattended, it is a common practice to inspect the installation after completion of the cycle and before the unit is put back on line. For many air filter systems, inspections between servicing periods are desirable. These inspections are intended to make certain that the units operate at optimum level.

Filter Life Estimating

A very important consideration when selecting an air filter is filter life. Longer filter life reduces filter cost per operating hour. Moreover, longer filter life results in less frequent filter changes which reduces annual labor and disposal costs.

Because of the many variables in the environment, a precise method for estimating filter life does not exist. Most data is based on user experience or manufacturers' life tests. Some effort has been made to correlate the life of a filter with its ASHRAE Dust Holding Capacity. However ANSI/ASHRAE Standard 52.1[(2)] specifically states:

> "Since the contaminant used for loading is not typical of natural atmospheric dust, its effect on the filter may not be the same as an equal amount of atmospheric dust. The value of the loading test is for rating and to some extent ranking filters. The performance values obtained in accordance with this standard cannot be used by themselves to predict the air cleanliness of a specific ventilated space or the service life of an installed filter."

Factors Affecting Filter Life

Although there is no reliable method for precisely calculating filter life, the life of a filter can be affected by a variety of factors including:

1. **Particle size distribution of contaminants**.

   Air composition is dependent on many different environmental variables, including geographic location and time of year. Additionally, the air that a filter system sees is dependent on the location of the air intake and source of the air in the system, whether it is outdoor air, recirculated air, or a mix of both. Simply put, no two air streams are exactly the same. Still air will normally contain the same relative percentages of different size dust particles. The total concentration may vary, but the relative distribution will remain constant. This fact is the basis for the classification of cleanrooms. On the other hand, turbulent atmospheric air may contain heavy concentrations of larger dust particles. An incorrectly located air inlet directly down stream of an incinerator stack may take in large quantities of fly ash. In each of these cases the variations in particle size distribution and concentration will affect filter life.

2. **Underrating or overrating of filters**.

   Filter rating refers to the amount of air that a clean filter will handle at a specified pressure drop. A typical statement is that "X filter is rated at 2000 cfm at 0.5 in. w.g. (0.944 m3/s at 125 Pa)." If the filter handles more air, the pressure drop will increase, usually in direct proportion to the amount of air being handled. Similarly, if it handles less air there will be a reduction in pressure drop. The effect of overrating and underrating an extended surface filter is important enough to deserve special consideration and is reviewed in greater depth in a later section of this chapter.

Using Prefilters to Extend Filter Life

A final filter's life may be extended by the use of a prefilter. A prefilter is a filter with lower overall efficiency installed in front of a higher efficiency final filter. The purpose of the prefilter is to capture the larger particles that block off air flow and cause a rise in the pressure drop of the final filter. This capture of larger particles extends the life of the final filter. Filter manufacturers' literature sometimes supplies information on the life extension provided by a prefilter.

Life extension of the final filter is the primary reason for using a prefilter. For this reason the owning and operating costs of the prefilter system must be less than the savings achieved by the life extension of the final filter. It should not be assumed that prefiltration is automatically the way to proceed. It is always desirable to run an owning and operating cost analysis of a system with and without prefilters to confirm that this assumption is correct. The operating cost must include the cost of power to overcome the resistance of the prefilter.

The one instance where prefilters should not be used is any application where toxic or hazardous dusts are involved. In these instances, all hazardous material should be collected in the final filter. The changing and disposal of a single filter loaded with hazardous material represents less of a risk to maintenance personnel than that of changing and disposing of several prefilters.

Underrating and Overrating Filters

A filter's rating and tested MERV refers to the amount of air that a clean filter will handle at a specified pressure drop. If the filter handles more air, the pressure drop will increase, usually in direct proportion to the amount of air being handled. Similarly, if the filter handles less air there will be a reduction in pressure drop. The filter's rated velocity is generally outlined in the manufacturer's literature. For example, most filters for use in commercial HVAC systems are rated at 2000 cfm/ 500 fpm velocity based on a 24" x 24" filter.

Example of Media Configuration and Service Life of 24" x 24" x 2" Pleated Filter:

Economy Capacity Pleated Filter =
8 square foot of media; usually changed 5 times per year

Standard Capacity Pleated Filter =
11 square feet of media; usually changed 4 times per year

High Capacity Pleated Filter =
17 square feet of media; usually changed 3 times per year

Overrating and underrating filters not only impacts the service life and pressure drop of the filter but it can also impact filter efficiency (see Chapter 2). Consult manufacturer recommended flow rates and MERV values before underrating or overrating any filter.

There are two reasons to operate an extended surface filter at a capacity lower than its rated airflow. One is extending the service life of a filter which makes it desirable to underrate a filter (operate it at less than its rated capacity) and two, the pressure drop usually decreases. These same two effects make it equally undesirable to overrate a filter (operate it more than its rated capacity). To understand this double effect, it is necessary to study one effect at a time. If clean air is flowing through a clean filter and the air flow is 80% of the standard rating, the pressure drop through the filter would also be 80% of its rated value. For example, if the rated pressure drop is 0.5 in. (125 Pa), the underrated pressure drop would be 80% of 0.5 in. or 0.4 in. (100Pa).

What is frequently overlooked is the fact that when a filter is underrated, there is a reduction in the rate of increase in pressure drop due to the fact that the rate of dust loading per square foot of media is also reduced.

Let us assume that at rated capacity the velocity of the air through a square foot of media in an extended surface filter such as a pocket filter is 25 fpm (0.127 m/s). If the rating is reduced to 80%, the velocity of air through the media will be 20 fpm (0.102 m/s). Let us also assume now that, instead of being clean, every cubic foot of air contains 1 unit of filterable dust (dust that will be captured by the filter). At rated capacity when the media velocity is 25 fpm, 25 units of dust will be captured every minute on each square foot of media. At the end of 10 minutes there will be 250 (10 x 25) more units of dust on a square foot of filter media. If, however, the media velocity is 20 fpm (0.102 m/s) the number of dust units deposited in a minute will be 20 per square foot. In 10 minutes the number deposited will be 200 (10 x 20) units of dust on a square foot of filter media. It will take 12.5 (10 x 250/200) minutes to deposit 250 units of dust.

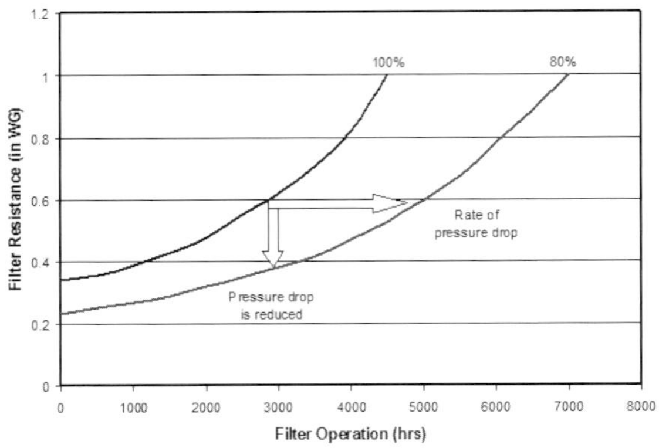

Figure 13.5.

To summarize, because of underrating, there will be an increase in filter life (the time required to reach a final pressure drop), for the following reasons:

1. Underrating reduces the pressure drop of the filter because the velocity of air through the filter medium is decreased.

2. The time required for a pressure drop increase due to captured dust will be extended. This is because the number of units of captured dust deposited on any unit area of media will occur more slowly when the filter is underrated.

These two points are illustrated in Figure 13.5 for an underrated filter.

The two effects are additive so that the life extension of a filter due to underrating is greater than would be expected by simply reducing the airflow and reducing the pressure drop. The reverse situation occurs when a filter is overrated. A set of typical life extension and reduction curves are shown in Figure 13.6.

Economics of Operating a Filter System at Other than Rated Capacity

There are several reasons for utilizing the double effect of operating an extended surface filter at other than rated capacity:

1. Underrating allows a filter to be operated to a lower final pressure drop while maintaining life equal to that at rated capacity.
2. Underrating allows a life extension so that a filter change might coincide with a factory shutdown or production changeover time.

Optimum Final Pressure Drop

In life cycle costing, filter life is important. Extended life can be attained by operating a filter to the highest designed final pressure drop that is acceptable. The longer the filter is in use, the lower will be the filter cost per unit of time. However, running the filter to a high final pressure drop may be more costly because the energy costs required to achieve this final pressure drop may use up all the savings in filter life extension.

A method[4] has been suggested to find the static pressure which optimizes the owning and operating cost of an extended surface filter. Graphically, it is shown in Figure 13.7 where

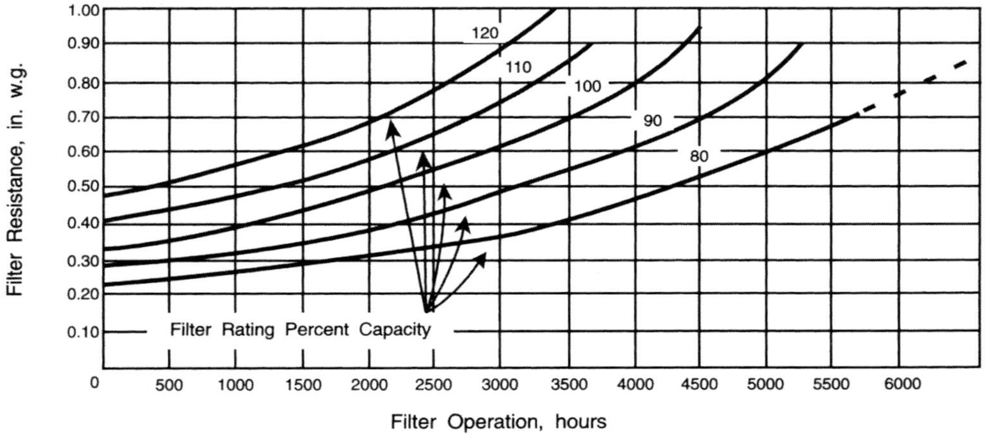

Figure 13.6. Effects of Underrating and Overrating on Filter Life.

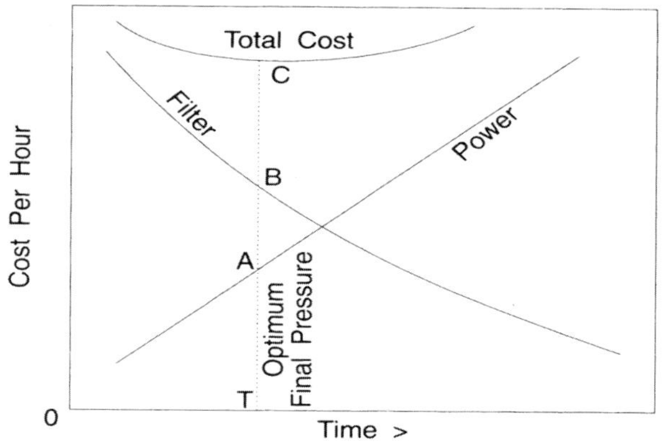

Figure 13.7.

both the filter cost per hour and the energy cost per hour are both plotted against time. Energy costs are low when the filter is clean but the filter cost per hour is high. The longer the filter is used, the lower its cost per hour becomes. At the same time as the filter resistance increases, the energy costs go up. At any point, the total cost per hour of operation is the sum of both the filter cost and the energy cost. The broken vertical line consists of two segments. The power cost is segment T-A. The filter cost is segment T-B. When they are added (segment A-C is the same length as segment T-B) they form line TC which represents the total cost pefpr hour of filter and power at time T. The curve for total cost vs. time starts high, hits a minimum, and rises again. The optimum pressure drop at which to change a filter is when the total cost is the lowest. This is at point C. Changing the filter before reaching point C is uneconomical because filter life is not optimized. Changing the filter after reaching point C is uneconomical because the cost of power is greater than the savings in extending the filter life.

Annual Owning and Operating Costs

The estimation of the annual owning and operating costs of a filter system consists of two parts:

1. **Annual Depreciation of Equipment**

   This is the annual share of the cost of the initial installation. For instance, if the total cost of the initial installation was $7,000 and the equipment was expected to last for seven years, the annual share of the installed cost would be $1,000 using the straight-line method of depreciation with no salvage value. This is the simplistic approach. In reality, tax considerations and accounting practices will play an important part in determining how much of first costs can be charged to annual owning and operating expense. These considerations and practices are too complex to detail here. However, ground rules should be established by all who will use an owning and operating cost analysis, detailing the accounting and taxation premises to be used in figuring this cost.

2. **Annual Operating Costs**

   The calculation of annual operating costs is much more straightforward since most operating costs are part of annual expenses.

   Annual operating costs are composed of:

   - Initial investment - first cost of filters
   - Inventory costs
   - Shipping Costs
   - Damage/Storage Loss Costs
   - Installation and Removal Costs
   - Disposal Costs
   - Energy Costs

   All costs should be included when determining filter cost.[6]

Filter Owning and Operating Cost Study Form

Much of the information about the owning and operating costs of a filter system can be summarized in a single form. Many have been designed for this purpose. Most are intended to allow a person to make a comparison of two or more filtering systems. A typical filter form is shown in Addendum 13.1, "Filter Owning and Operating Cost Worksheet."

13.8

All of the input data is needed to produce the output of life cycle costing. Fully loaded labor costs should include cost per hour plus benefits and other labor costs.

Fan Blower/Drive efficiency may be obtained from manufacturer data or estimated at 80%

Notes to Chapter 13

1. Carlsson, Thomas; "Indoor Air Filtration: Why Use Polymer Based Filter Media", Filtration+Separation, Volume 38 #2, March 2001, pp 30-32.
2. ANSI/ASHRAE Standard 52.1 - 1992 Gravimetric and Dust Spot Procedures for Testing Air Cleaning Devices Used in General Ventilation for Removing Particulate Matter, pp.1 American Society of Heating, Refrigerating and Air-Conditioning Engineers Inc., Atlanta GA.
3. Tips on Energy Conservation (Publication EC-1) American Air Filter Co. (1974).
4. R.H. Avery Air filtration: resistances, energy, and service life, H/P/A/C December 1973.
5. R.H. Avery, Energy Effective Air Filtration, Plant Engineering (Sept. 28, 1978).
6. Air Filtration Association, 2004.

Addendum 13.1

Life Cycle Cost Analysis Tool

Input Data

		Filter A	Filter B
1	Filter Type		
2	Filter Model		
3	Model #		
4	Filter Price ($ per filter)		
5	Number of Filters Per Case		
6	Shipping Cost Per Case ($)		
7	Estimated Damage Loss (%)		
8	Number of Filters in Bank		
9	Estimated Filter Life (months)		
10	Changeout time required = full bank (min)		
11	Changeout Labor Cost ($/hour fully loaded)		
12	Disposal Cost ($/filter)		
13	Initial Resistance ("WG)		
14	Recommended Final Resistance ("WG)		
15	System Airflow Rate (cfm)		
16	Days in Operation Per Year		
17	Hours in Operation Per Day		
18	Energy Cost ($/kWH)		
19	Fan/Blower/Drive Efficiency (%) = Actual % or Use 80%		

OUTPUT DATA

Initial Investment Costs			
20	Number of Filters = Line 8		
21	Filter Price = Line 4		
22	Estimated Filter Life (months) = Line 9		
23	Number of Changeouts/Year- 12 ÷ Line 22		
24	**Subtotal Annual Filter Costs-Line20xLine21xLine23**		

Addendum 13.1.2

		Filter A	Filter B
Inventory Costs			
25	# Filters Used/Year = Line 20 x Line 23		
26	# Filters/Case		
27	Number of Cases Used/Year Line 25 ÷ Line 26		
28	Actual # Cases Purchased = Roundup for Odd Numbers		
29	"Extra" Filters Purchased/Year		
30	Filter Cost = Line 21		
31	**Subtotal Annual Inventory Cost Line 29 x Line 30**		

Shipping Costs			
32	Shipping Cost/Case = Line 6		
33	# cases Purchased/Year = Line 28		
34	**Subtotal Annual Shipping/Storage Costs = Line 32xLine 33**		

Damage/Storage Loss			
35	Estimated % Damage Loss = Line 7		
36	# filters Used/Year = Line 25		
37	Cost/Filter = Line 30		
38	**Subtotal Annual Damage/Storage Loss = Line 37xLine36xLine35**		

Installation/Removal Costs = Full Cycle			
39	Time Required/Changeout (minutes) = Line 10		
40	Time Required/Changeout (hours) = Line 39 ÷ 60		
41	# Changeouts/Year = Line 23		
42	Fully Loaded Labor Cost = Line 11		
43	**Subtotal Annual Installation/Removal Costs = Line 42xLine40xLine 41**		

Disposal Costs			
44	Disposal Cost/Filter = Line 12		
45	# Filters Disposed/Year = Line 25		
46	**Subtotal Annual Disposal Costs = Line 45xLine 44**		

		Filter A	Filter B
Energy Costs			
47	Initial Resistance (Pa) = Line 13x249.09		
48	Recommended Final Resistance (Pa) = Line 14x249.09		
49	Average Resistance (Pa) = Line 47+Line 48 ÷ 2		
50	System Airflow (m3/sec) = Line 15 ÷ 2120		
51	Filter Airflow (m3/sec) = Line 50 ÷ Line 8		
52	Filter Life = (Line 16xLine 17) ÷ (12 ÷ Line 9)		
53	Energy Consumption (kwh) = (Line 49xLine 51xLine 52 ÷ 0.8) ÷ 1000		
54	Energy Cost Per Filter ($) = Line 53xLine 18		
55	Energy Cost Per Changeout ($) Line 54xLine 8		
56	**Subtotal Annual Energy Cost ($) = Line 55x12 ÷ Line 9**		

TOTAL COSTS			
57	Annual Filter Costs = Line 24		
58	Annual Inventory Costs = Line 31		
59	Annual Shipping/Storage Costs = Line 34		
60	Annual Damage/Storage Loss = Line 38		
61	Annual Installation/Removal Costs = Line 43		
62	Annual Disposal Costs = Line 46		
63	Annual Energy Costs = Line 56		
64	**Total Life Cycle Cost = Add Lines 57 = 63**		
Comparison Between Filters			

Customer Name: ______________________________

Customer Address: ______________________________

Prepared By: ______________________________

Date: ____________________

14

Ultraviolet Germicidal Irradiation, Photocatalytic Oxidation

New forms of air cleaning technology are continually being developed and older technologies are being refined and utilized to clean indoor air of contaminants. While this "cleaning" may not incorporate capture of particulate, other established technologies can affect certain particulates and odors in positive ways when applied properly.

Two of the more popular technologies that have been proven over time and used to clean the indoor environment are ultraviolet light in the "C" wavelength, known as ultraviolet germicidal irradiation (UVGI or UVC) and photo catalytic oxidation (PCO).

Ultraviolet Germicidal Irradiation - UVGI

Short-wave ultraviolet radiation in the "C" band, or UVC (200 to 280 nanometer wavelength, see Figure 14.1) has been used since the 1800's to inactivate all types of micro-organisms: mold (fungi), bacteria, virus, and yeast. Niels Ryberg Finsen (1860-1904) was the first to employ germicidal UV rays while researching treatment of skin diseases. In the early 1900's UVGI was first utilized to disinfect water in Marseilles, France and in the 1930's, Westinghouse was the first to develop a commercial grade UVC lamp that was used primarily in hospitals for upper-room air irradiation where air currents moved pathogens past the germicidal UV lamps.

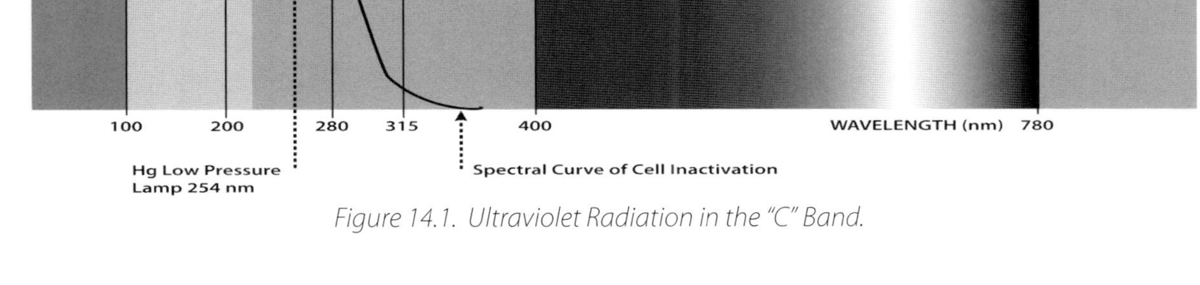

Figure 14.1. Ultraviolet Radiation in the "C" Band.

Ultraviolet light is just below violet, which is the lowest wavelength of visible light in the electromagnetic spectrum. The various UV wavelengths produce different effects. UVA is also known as black light. UVB is used in tanning booths to produce a suntan, UV in the C Band, specifically at 265 nanometers, has a germicidal effect on microorganisms. UV wavelengths from 100 to 240 nanometers produce some ozone with the 185 nm wavelength producing peak levels of ozone. Many, but not all, UVC lamps are shielded from producing ozone.

In the 1950's, UVGI was successfully used to control the spread of tuberculosis in hospitals with the use of "upper room air" irradiation. As mechanical ventilation systems developed, UVC lights were placed in air ducts with little success. This was because the ballasts and lamps being made at the time did not perform well in cold, moving air. Further, the UV dose, the amount of irradiation energy and exposure time of a microorganism to the UVC energy needed for inactivation, was limited. These factors all had an adverse effect on UVGI performance.

In 1995, commercial research teams introduced their work in updating the technology and establishing the effectiveness of UVGI in a moving air stream. UVC lights were placed at the cooling coil, where velocities are lower than in ducts (longer exposure time) and where microbial activity is known to occur. In addition, aluminum, the most common fin stock used in cooling coils, is highly reflective of UVC light. There are now many companies offering a variety of UVGI systems for HVAC applications.

In addition, computer software has been developed that enables prediction of surface and air stream disinfection rates on a variety of common microorganisms. See Figure 14.2. This is especially important in hospitals and clinics where certain airborne micro-organisms (also known as bio-aerosols) such as tuberculosis are targeted. Further developments in disinfection modeling software are ongoing.

How UVGI Works

The DNA of the microorganism absorbs UVC energy. Absorption of a sufficient amount of UVGI, referred to as UV dose or fluence, causes molecular instability which results in DNA strand breaks and lesions. The cell is no longer able to grow or reproduce. At very high levels of exposure to UVC (increased exposure time and/or high intensity), the molecules of the microorganism begin to break down as the

Microorganism	Type	UV Dose (mJ/cm^2) per Log Reduction of:						
		1	2	3	4	5	6	7
Aeromonas hydrophila ATCC7966	Bacteria	1.1	2.6	3.9	5	6.7	8.6	
Campylobacter jejuni ATCC 43429	Bacteria	1.6	3.4	4	4.6	5.9		
Escherichia coli ATCC 11229	Bacteria	2.5	3	3.5	5	10	15	
Legionella pneumophila ATCC 43660	Bacteria	3.1	5	6.9	9.4			
Salmonella anatum (from human feces)	Bacteria	7.5	12	15				
Shigella sonnei ATCC9290	Bacteria	3.2	4.9	6.5	8.2			
Staphylococcus aureus ATCC25923	Bacteria	3.9	5.4	6.5	10.4			
Streptococcus faecalis ATCC29212	Bacteria	6.6	8.8	9.9	11.2			
Bacillus subtilis spores ATCC 6633	Spores	29	40	51				
B40-8 Phage	Phage	12	18	23	28			
Adenovirus 41 ATCC TAK	Virus	22.4	49.5	80.2				
Coxsackievirus B5	Virus	6.9	13.7	20.6				
Reovirus-3	Virus	11.2	22.4					
Rotavirus SA-11	Virus	7.6	15.3	23				

Figure 14.2. The Chart Shows UV Dose or Fluence Required for Log Reductions of Specific Microorganisms.

microorganism is degraded into carbon dioxide and water. In virus particles, which need a "host" to support their life cycle, the RNA is affected similar to the DNA in living microorganisms.

As shown in Figure 14.2, UV dose or fluence depends on the type of microorganism. In a study of UVGI air stream disinfection done at Research Triangle Institute in 2002, it was observed that "a substantial inactivation of airborne environmental fungi was accomplished with UVC dose levels readily achievable using multiple lamps. The vegetative bacteria tested were relatively easy to inactivate, while the bacterial spores tested displayed an intermediate response. Fungal spores were difficult to inactivate." [1]

UVGI and HVAC Cooling Coils

Poorly filtered coils will catch and hold inert and biological particulate matter over time. While the particulate collected on heating coils does have some insulating effect, the result on cooling coils is a buildup of bio-film on and in between the fins and in the drain pan of the HVAC unit. This is due to the coils function of removing latent heat (the heat associated with water vapor) from the air and producing condensate. This bio-film is comprised of particulate and microbial growth, primarily mold growth in the coils and bacterial growth in drain pan condensate. Bio-film impairs the heat transfer efficiency of the coil and increases the velocity of air through the coil which further typically degrades the coils ability to remove heat. HVAC coils are located downstream of air filters. Contaminants and their by-products (odors, allergens, etc...) can be entrained into the supply air and onto supply duct insulation.

When the coil is exposed to a properly sized UVGI light system, the germicidal effect of UVC on microorganisms and the washing effect of the condensate combine to clean organic matter throughout the depth of the coil over a period of time. See Figure 14.3. UVGI is normally placed on the downstream side of the coil and over the drain pan. The result is restored or maintained heat transfer efficiency, reduced coil pressure or increased air volume, and improved IAQ. If the AHU blower and/or the chiller system have variable frequency drives (VFD's), energy savings from reduced horsepower are possible as the UVGI lights clean the cooling coil. When UVC lights are placed over the drain pan, the need for pan treatments is eliminated, resulting in additional direct cost savings.

Other UVGI Applications in HVAC

In addition to cooling coil irradiation, UVGI is also used for air stream disinfection in HVAC air ducts to reduce specific microorganisms to reduced levels. Another emerging application is the use of a specific UVR (UVGI Rating Value) UV light system in combination with a particulate filter. This combination enables air stream disinfection of UVC susceptible bio-aerosols and capture of UV resistant microorganisms by the air filter. This system is referred to as "Capture and Kill." This concept also holds promise for reducing infectious particles on final filters in hospital and laboratory applications.

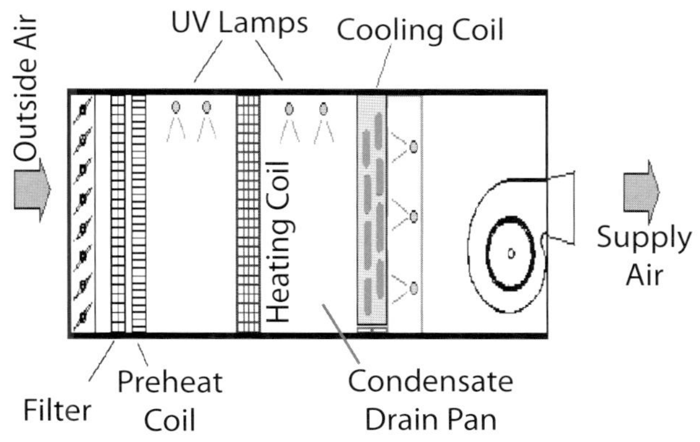

Figure 14.3. Typical UV Light Placement in HVAC System

UVC and Safety Issues

UVGI acts on any organic matter, which is anything not mineral in nature. Exposure to UVC light will adversely affect the skin and eyes. While UVC does not penetrate deep enough to cause lasting damage, the primary visual hazard for UVC is conjunctivitis which produces an irritation of the mucous membrane of the eyes. The result of skin exposure to UVC is erythema which is an irritation of the skin similar to sunburn. These symptoms usually recede within 24 to 36 hours after exposure. To avoid these problems, UVC lights should be deactivated before servicing. It is recommended in most UV light specifications that a safety interlock be installed on access doors. If this is not possible, goggles should be worn and exposed skin protected.

UVGI can also cause the breakdown of certain types of organic materials used in HVAC systems including open cell foam and polypropylene fibers such as those used in passive electrostatic filters and electrostatically charged synthetic filter media. Thin, clear plastics can become brittle with prolonged exposure to UVC. These materials should be protected from direct exposure to UVC light. Pigments in plastics, especially carbon (used in black plastics) absorb the UVGI energy which results in little or no damage, this includes the pigments and binders used in many synthetic and fiberglass filter media. Of course, the manufacturer of the product in question should always be consulted regarding exposure of UVC to their product.

Unlike UVA and UVB, UVC has very little penetrating power. It will not penetrate through regular glass, plastic, or most other solid materials.

UVGI System Operation

UVGI lights are similar in operation to fluorescents with electronic ballasts powering low pressure mercury vapor lamps. Although they produce some visible light, most of the energy emitted is UVC at 254 nanometers, very close (85%) to the optimum 265 nm germicidal wavelength. There are a wide variety of UVC lamps and fixture configurations available.

"There is, as yet, no conclusive evidence demonstrating that any one type of lamp is more effective against the broad range of pathogens and allergens in the indoor environment. Therefore, the selection of which type of lamp to use will depend on factors other than disinfection performance, such as energy consumption, lamp cooling effects, ease of installation, etc..."[2]

Each fixture manufacturer provides sizing and placement recommendations for their products. Normal lamp replacement is between one and two years and ballasts are typically rated at 50,000 hours of service life.[3]

UVGI lamps lose output intensity in cold, moving air. Quartz and PTFE sleeves are used to stabilize output in these conditions. The polymer sleeves are also used to encapsulate the lamp quartz in the event of lamp breakage. Reflective materials, especially aluminum, are also utilized to increase the amount of UVC energy, enhancing the germicidal effect.

UVGI Applications and Standards

ASHRAE, ARI (Air Conditioning & Refrigeration Institute), and IUVA (International Ultra Violet Association) have coordinated efforts to develop testing standards and guidelines for use of UVC technology in various applications.

IUVA has developed application guidelines and recommendations for installation and testing of UVC light systems. Several have been released for public comment and more are pending release. These documents are available at www.iuva.org and serve as source documents for ASHRAE and ARI. ARI's focus is on developing rating standards and field performance verification. ASHRAE TC 2.9 has its subcommittees developing test and installation standards. The 2008 HVAC Applications Handbook will have a chapter devoted to UVGI. All three organizations recognize the value of this technology as UVC light installations become more commonplace. As with air filters, fans, motors, and other HVAC components, standards and application guidelines protect the end user and serve to advance the technology as manufacturers compete to meet or exceed the industry standard.

Photocatalytic Oxidation (PCO)

Photocatalytic oxidation utilizes ultraviolet or near-ultraviolet (< 385 nm) radiation to promote electrons from the valence band into the conduction band of titanium dioxide, zirconium dioxide, or an aluminum oxide semiconductor. Sometimes the catalyst is a combination of several of these. Destruction of organic compounds takes place through reactions with molecular oxygen or through reactions with hydroxyl radicals and super-oxide ions formed after the initial production of highly reactive electron and hole pairs. See Figure 14.4.

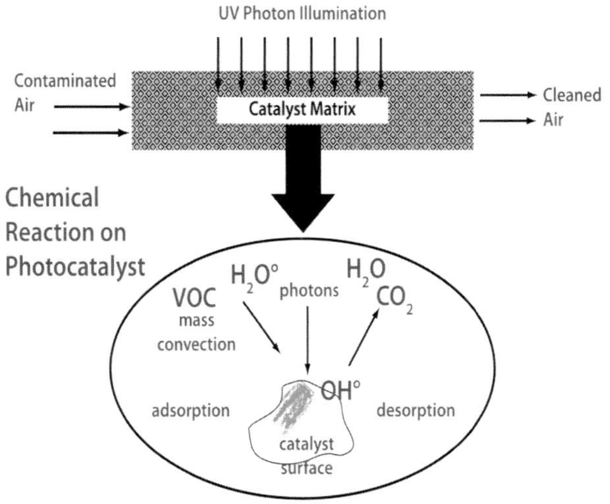

Figure 14.4. Illustration of PCO Process

PCO provides significant advantages for the mitigation of pollutants associated with poor indoor air quality. Due to low pressure drop across the reactor and ambient temperature operation, PCO reactors can be incorporated into existing HVAC systems. Rather than transferring pollutants from the gas phase to the solid phase, PCO provides a reduction of absolute toxicity as the gaseous products from the complete photocatalytic oxidation of volatile organic compounds are carbon dioxide and water.

There are four typical PCO reactor designs. They are: the packed bed annular reactor, the coated wall annular reactor, the box reactor and the flat bed PCO reactor.

Coated-Wall Annular reactor is coated on the inside surface of its outer glass tube with the titanium dioxide photocatalyst. This design allows for uniform light distribution and a catalyst surface that runs the length of the reactor body. See Figure 14.5.

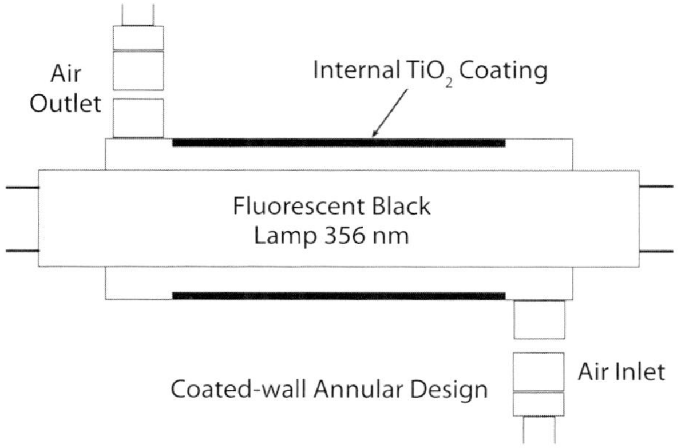

Figure 14.5. Coated-Wall Annular reactor

Packed Bed Annular reactor is similar in design to the coated-wall annular except it has greater surface exposure area since it utilized 3 mm glass beads coated with titanium dioxide. See Figure 14.6.

Box reactor has the largest reaction surface

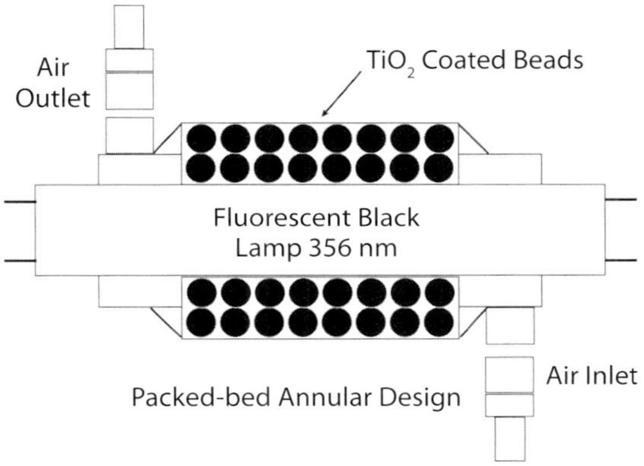

Figure 14.6. Packed Bed Annular reactor

area provided by two flat panel fiberglass filters coated with titanium dioxide placed on either side of the ultraviolet light. See Figure 14.7.

Flat Bed reactors use the flat panel fiberglass

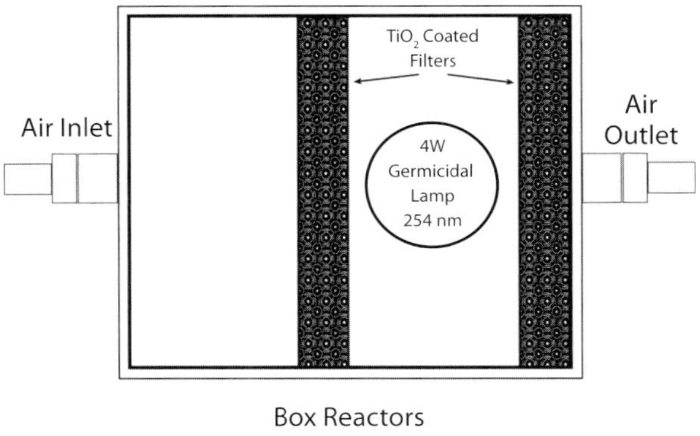

Figure 14.7. Box reactor

filter under a flat quartz glass pane that allows for uniform light exposure and external variations in UV light intensity. See Figure 14.8.

The design objectives for any photocatalytic reactor are to mix photons, catalyst and reactants as intimately as possible. Control of UV light intensity is also desirable.

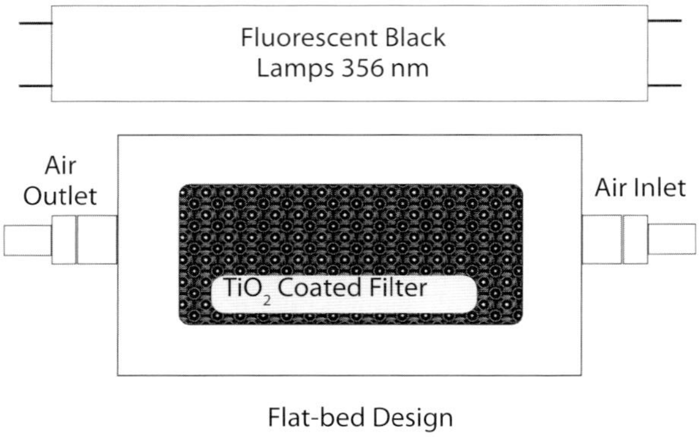

Figure 14.8. Flat Bed reactor

PCO has a wide variety of applications. The application with the best chance for immediate application is the destruction of volatile organic compounds (VOC) in the air. The PCO process can be applied directly to airborne pollutants as well as to gaseous pollutants released during certain remediation techniques such as removing chemicals from contaminated soil.

PCO has several advantages for destruction of VOC's such as a low pressure drop through the reaction chamber, it can be incorporated into HVAC systems, it works with very low concentrations of contaminants, and it can take place at ambient or near ambient temperatures. In most instances, the resultant product of PCO is CO_2 and H_2O (carbon dioxide and water vapor).

A disadvantage of PCO can be the high cost of the equipment relative to other technologies such as activated carbon. This disadvantage can be offset by comparison to other technologies that require destruction of VOC's or to the higher cost of not being able to effectively remove certain VOC's from the air.

Notes to Chapter 14

1. "USE OF IN-DUCT UVC LAMPS TO INACTIVATE AIRBORNE ENVIRONMENTAL BACTERIA AND FUNGI," D.W. VanOsdell and K.K. Foard, Proceedings: Indoor Air 2002
2. W.J. Kowalski, PE, PhD
3. Lighting Industry Federation Technical Statement #34, September 2004

15

Industrial Contaminant Air Filtration Control Systems

Editor's Note: *The information in this chapter is not meant to replace or override any guidelines or standards established by OSHA, NIOSH, ACGIH or any other government body having authority over industrial workplaces. This chapter is for information purposes to be used as a general overview on industrial contaminant air filtration. NAFA encourages users to ensure they are in compliance with all standards and recommended exposure limits established by cognizant authorities.*

Metal Working Fluids

The metal working industry requires the use of various Metal Working Fluids (MWF) in their production to cool, lubricate and clean the produced parts and the tools. MWFs are used in many operations such as drilling, cutting & boring. (See Figure 15.1.) In these operations, MWFs perform many functions including flushing debris away from the cutting operation, reducing the temperature of the cutting tools (thereby improving cutting tolerances and prolong cutting tool life), and reducing the temperature of the production part.

All of these functions are important. In the automotive industry manufacturing tolerances are extremely tight, often requiring precise measurements with tolerances measured by the thousands of an inch. In these tight-tolerance cases, the function of reducing temperature is extremely important. During modern metalworking operations, cutting tool pressure exceeding 400,000 pounds per square inch and rotational speeds of over 20,000 rpm can cause temperatures of the cutting tool and item being worked to exceed 1400 degrees Fahrenheit. Without lubrication and cooling from MWFs, these extreme temperatures can damage the cutting tools and the material being cut, bored or drilled.

Additionally, the material can expand in size while heated then contract back to its original size when it cools. If this occurs on a part which requires an exact distance between three holes, then a change in that distance may affect the installation of the part, or in the worst case, recalls due to defective parts.

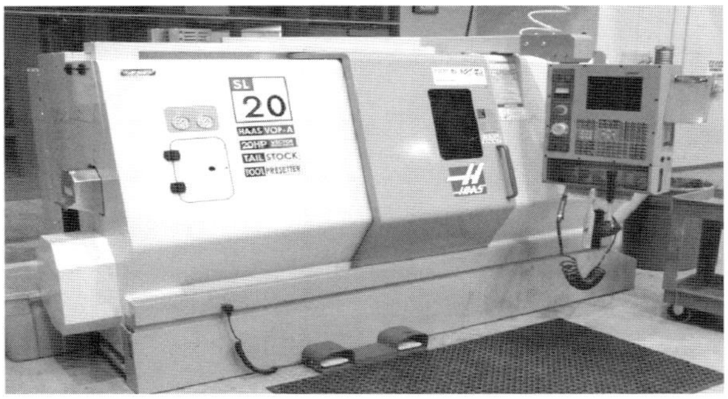

Figure 15.1. A modern metal machining station.

Types of Metal Working Fluids

The composition of these MWFs has changed over the last few decades from primarily animal-based oils to synthetic and or glycol-based oils containing varying percentages of water, oils, surfactants, and other ingredients. The types of MWF are too varied to list, but can be summarized into four types:

1. **Straight Oil** (neat oil) are severely refined solvent-refined petroleum oils (lubricant-based oils) or other animal, marine, vegetable, or synthetic oils used singularly or in combination and with or without additives. Straight oils are not designed to be diluted with water.
2. **Soluble oil** (emulsifiable oil) are combinations of 30% to 85% severely refined lubricant-based oils and emulsifiers that may include other performance additives. Soluble oils are diluted with water at ratios of 1 part concentrated to 5-40 parts of water.
3. **Semi-synthetic Oil** are metal working fluids contain a lower amount of severely refined lubricate-based oil in the concentrate (5% to 30%), a higher proportion of emulsifiers , and 30% to 50% water. The transparent concentrate is diluted 10 to 40 parts water.
4. **Synthetic Metal Working Fluids** contain no petroleum oil and may be water soluble or water dispersible. The synthetic concentrate is diluted with 10 to 40 parts water.[1]

Oil Mist and Oil Smoke

Oil mist and oil smoke particles are formed in three different ways:

1. High pressure spraying. When MWF is sprayed with high pressure the liquid will form mist or an aerosol. A higher pressure spray will form smaller particles compared to a lower pressure.
2. High speed/rotation. When the MWF hits a rotating cutting head or tool some of it will be thrown off as an oil mist. High-speed production forms smaller mist particles compared lower speeds/rotation.
3. High temperature. A work piece or a tool with high temperature will boil or burn the MWF, this will cause an oil smoke, or a combination of smoke and mist.

It should be noted that oil mists and oil smoke are often collectively referred to as "oil mist", however they differ significantly in the contaminant particle size. Oil mist is generally larger airborne particles of MWF (1.0 – 10.0

	Oil Mist	Oil Smoke
Particle Size	1.0 to 10.0 micron	< 1.0 micron
Settling Rate	10.0 micron – 7.2 inches per minute 1.0 micron – 0.1 inches per minute	0.5 micron -- 1.4 inches per hour 0.1 micron – 1.14 inches per day
Respirability	3.0 – 10.0 micron -- Less respirable 1.0 – 3.0 micron – more respirable	Very respirable
Volume	Large Volume	Small Volume
Collection Challenge	Large volume can overwhelm certain filtration systems Can contain solid particles such as fine metal and/or tool-head fragments	Small size will pass through many types of filters, yet still rapidly plug or foul a high efficiency filter

Figure 15.2

micron in size), while oil smoke is made up of smaller particles (below 1.0 micron). These two types of airborne contaminant can pose differing risks and require different control techniques. A summary of these differences is included in the Figure 15.2.

Health and Safety Risks

Health risks associated with oil mist and oil smoke include either dermal (such as dermatitis and oil acne), respiratory (such as asthma, pneumonia) or systemic (cancer). Oil mists also may contain particulate from the metal being cut and tool wear and bacteria & fungi which develop in the fluid.

In 1992, the Scandinavian Journal of Work, Environment & Health published a study which reported that metalworkers faced a much higher risk than then in general population for contracting various kinds of cancers. At the time, the National Institute for Occupational Safety and Health (NIOSH) had a standard permitting a maximum allowable concentration of oil mist of 5 mg/M^3. The United Auto Workers stepped in and filed suit in protection of its workers, requesting OSHA to reduce that level immediately. The UAW felt that lowering the NIOSH level would result in increased worker health and productivity, and reduce the amount of respiratory diseases in their metalworkers. NIOSH agreed with the UAW, paving the way for the standard as it exists today.[1]

In March of 1998, NIOSH published their findings in a document entitled, *"Criteria For A Recommended Standard OCCUPATIONAL EXPOSURE TO METALWORKING FLUIDS,"* DHHS (NIOSH) Publication No. 98-116. This document explains the various hazards which exists in the metal working industry and also makes recommendations to reduce the possibilities of worker endangerment. The recommendation by NIOSH is that the older recommended exposure limit (REL) of 5 mg/M^3 was too high. The new REL is now 0.4 mg/M^3 for thoracic particulate mass which is that amount of aerosol which bypasses the larynx within the respiratory system. Since thoracic samplers were relatively unavailable, NIOSH allows a measurement of total particulate mass level as a substitute, using 0.5 mg/M^3 as the REL as a time weighted average concentration for up to 10 hours a day in a 40 hour work week. It is noted that NIOSH recommends that owners should attempt to reduce the WMF mist concentration below the REL "whenever possible" since some workers can develop work-related asthma or hypersensitivity pneumonitis at levels below the new NIOSH REL. Furthermore, since past exposures to MWF mist has been linked to some cancers, it is important to limit the exposure level in general.

In addition to changing the REL from 5 mg/M^3 to 0.5 mg/M^3, NIOSH has other recommendations & control measures to help reduce MWF mist-associated health issues. Some of these are listed below.

- Safety and health training to identify hazardous situations and keep employees, visitors, and on-site contractors informed
- Work site analysis to include routine monitoring of employees
- Medical monitoring of exposed workers
- Hazard prevention and control which includes:
  - Proper fluid selection and design and operation of the fluid delivery system.
  - Use of oil mist collection systems properly designed for the operation and the use of proper machine enclosures to contain the airborne mist. This is the area with which NAFA has the most direct involvement and will be covered in detail below
  - Using guards on coolant return trenches to prevent entrance of tramp oils
  - Proper fluid maintenance
  - Proper use of biocides in coolant to avoid fouling
  - Proper isolation of mist-generating situations
  - Use of ventilation systems
  - Use of protective clothing and equipment
  - Sanitation and hygiene
  - Labelling and posting of the existence of hazardous substances

For a complete list of these recommendations, NAFA recommends end users consult DHHS Publication No. 98-116.

Oil Mist and Smoke Collection

It is generally understood that is better to capture and filter a contaminant at the source ("source capture") than it is to exhaust the contaminant or allow it to escape into the workplace environment and then try to filter it in the general airspace ("general air cleaning"). Exhausting of oil mist and smoke causes building air imbalances (requiring make-up air systems and their associated expense), can be detrimental to facility roof materials and other equipment, pooling of oil at the exhaust point (see Figure 15.3), and environmental pollution. General air cleaning, to be effective, requires 10-25 air changes per hour throughout the space being cleaned, which is a much higher airflow than required in a source capture air cleaning system.

Oil mist and oil smoke contain large amounts of aerosols (several times higher than common polluted air) and the aerosols are both liquid and solid, and very small (typical size is 0.1-5.0 µm). These are the challenging factors that make oil mist very difficult for simple filters or general-use filtration systems. Most standard air filters, designed to filter atmospheric air, cannot handle the large volumes generated in an oils mist application and can degrade significantly when placed in a wet environment. Standard dust separators, on the other hand, can handle the volume of contaminant, but rely on a self-cleaning or "pulsing" mechanism that is rendered useless in an oil mist application.

Figure 15.3. Exhausting of oil mist causes pooling of oil at the exhaust point.

Therefore, many differing technologies have been developed specifically to overcome the challenges of oil mist and oil smoke collection, while trying to provide a cost-effective, yet highly efficient method to control mist levels at or below the OSHA / NIOSH recommended exposure limits.

System Performance Goals

The recommended performance features of a properly designed source-capture mist collection system are as follows:

- **Continuous airflow rates at design:** To efficiently pick up the mist and smoke at the hood point or properly evacuate the machining enclosure on a continuous basis without plugging or falling off in airflow.
- **Continuous high filtration efficiency:** Incorporate the proper combination of mist & smoke removal mechanisms to provide air at the outlet with continuous mist levels of 0.20 mg/m^3 or less. Some filtration can degrade in operation to an efficiency far below the original rating.
- **Low maintenance:** A good source-capture oil mist collection ***system*** is a combination of the following parts:
  - Machine enclosure, hood or suction point
  - Ducting
  - Oil mist & smoke separator section (or sections)
  - Blower and motor

To get good working filtration system all parts listed above need to be properly designed to work in concert. While the machine enclosure, ducting and blower/motor components are critical, they are beyond the scope of this document. The focus of the remainder of this document will be on the oil mist and smoke separators available.

Oil Mist and Smoke Separator Systems

As mentioned, several technologies have been developed specifically for the removal of oil mist and oil smoke. These technologies can be divided into three main types:

1. Media-type filtration systems
2. Centrifugal separators or cyclone separators
3. Electrostatic precipitators

Each of the above systems may have distinctly different performance characteristics in differing applications. Differing MWF used, differing head rotation speeds and cutting temperatures, and different enclosures are just a few of the variables that may prevent or enhance the performance of the mist separator. In addition, the combination of several of the above technologies can often provide the best performance and cost-effectiveness. For example, a HEPA filter (Media-type) must be combined with the other technologies acting as prefilters so as to not plug quickly and cause high maintenance and high cost.

1. Media-type filtration systems

Media-Type filtration systems are usually designed as progressive systems starting with prefiltration and progressing to higher efficiencies. The first stage is usually what is called a tortuous path, meaning the design of the air intake on the separator forces the air to drastically change its direction, normally from horizontal to downward vertically. Then, the air typically turns 180 degrees and is forced up through the first stage of filtration. This rapid change in airflow direction causes the large droplets to maintain their momentum, and not follow the air stream (impingement) and drain down the filter into a sump. This filter is usually a metal baffle or expanded metal type filter.

The air then travels from the baffle section upwards through the next two or three stages of filtration, depending upon the design. After the metal filter, the air will then go through a higher efficiency filter, either a 12" deep rigid filter, or a specially designed "oil mist" pocket filter. Much of the finer mist is collected in this stage, and forms into larger droplets called coalescence, which fall back against the flow of air into a coolant collection drain pan. Filter medias are hydrophobic/oliophobic, because they do not absorb moisture vapors, have lower pressure drop restrictions and generally lead to longer filter life and higher performance. In units using pocket filters, they normally include either one or two loops sewn to each pocket to hang the pockets vertically within the separator. This is important for two reasons: 1) to keep the pockets vertical which allows for the collected mist to drain back out of the pockets, and 2) to allow the machine to be shut off and turned back on without the pockets falling from the weight of the collected oil. The final stage in these separator units is usually a 95% DOP or 99.97% HEPA filter, depending upon the manufacturer, the MWF and the application. Some units also incorporate molecular filtration to collect gas phase contaminants. This final stage ensures compliance with the NIOSH REL by providing air which can then be re-circulated back into the plant.

Advantages: High constant separation efficiency

Disadvantages: The service life of a pocket filter will decrease if it is connected to application with high levels of oil mist or if the mist contain large amounts of small particles.

Another type, cartridge oil mist separators, use the pleat technique to get more filter area at the same space. These separators usually have a labyrinth and/or a metal mesh type pre-filter that capture the largest particles. The cartridges have a design with a thin filter media made of fibers that allow the oil drops to coagulate and drain out of the filter. Progressive filtration of hydrophobic/oliophobic materials that do not adsorb moisture vapors and provide a natural moisture barrier is recommended. Some separators have a HEPA filter as the last filter stage as standard or as required.

Advantages: Compact design

Disadvantages: Design configuration of filter parts.

2. Centrifugal Separators

These separators use centrifugal force to separate out particles from an airstream. The force is created by spinning the air with a cone-shaped plenum, spinning a filtered drum within a plenum, or spinning a mechanism such as a disc-stack within a plenum to force particles out of an airstream. Inertial deposition of solid and liquid particles is increased with higher airspeed/centrifugal force. The techniques used in centrifugal separators are cyclones, rotating filters and/or rotating disc-stack separators.

Cyclones force the air to spin at a high rate by forcing air into a cone-shaped plenum. This causes centrifugal force which forces the mist to leave the air stream and collect on the sides of the cyclone. (See Figure 15.4.)

Advantages: Simple design, no replaceable filters.

Disadvantages: Will only capture big and/or heavy particles.

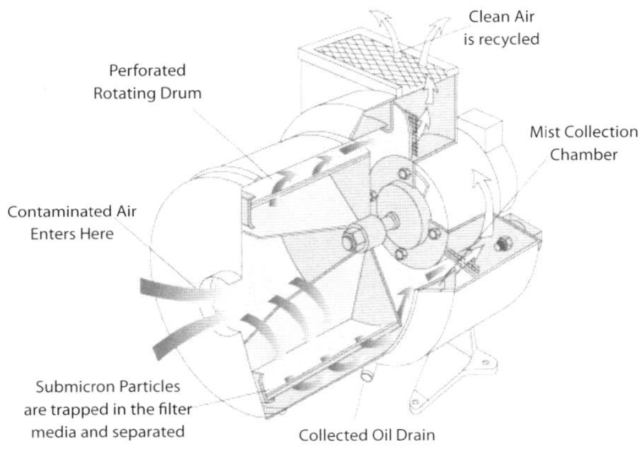

Figure 15.4. Typical Centrifugal Separator.

Rotating filters have an inlet in the center of a rotating, filtered drum. The rotation of the drum forces the air through a filter media that catches and coagulates liquid particles. The now larger particles then are forced to the outer wall of the separator and drain. An after filter is often used to capture the smaller particles that pass through the rotating filter.

Advantages: Low cost and compact design.

Disadvantages: Low separation efficiency for small particles, especially particles sized 1 micron or smaller. Sensitive to imbalance caused by chips, sludge and solid particles.

Rotating or Disk-Stack separators have a center inlet in the bottom part of the unit. The polluted air is sucked in to the center of the separator. The separator is a disk stack with a high number of plastic disks stacked on top of each other with a small distance between them. The rotation of the disk stack forces the air between the disks at an extremely high velocity. The particles leave the disk stack with a high force of gravity and hit the outer wall of the separator, where they coagulate and drain. Some rotating disk-stack separators have a HEPA filter as a last filter stage as standard or as an accessory.

Advantages: Compact design, no frequent filter change-out, high separation efficiency.

Disadvantages: Sensitive to chips, sludge and solid particles.

3. Electrostatic precipitators

When a particle is electrically charged it can be attracted and captured on an oppositely charged plate in the collector section. These electrostatic precipitators (called ESPs or electronic air cleaners), were invented in the early 1900s. In the 1970s these precipitators became a common method to clean oil mist from an airstream. (See Chapter 6 and Figure 15.5.)

Electrostatic precipitators typically have a five part design:

1. A baffle and/or a metal mesh type pre-filter that capture and drain the biggest particles.
2. An ionizing section that applies an electrical charge to the particles.
3. A collecting section with oppositely charged collection plates. The particles, charged in the ionizing section, are attracted and captured to collection plates and drained back to the source (machine).
4. A fan that transport the air.
5. An after filter (often a HEPA filter).

When an electrostatic precipitator is connected to machines that produce a lot of oil mist it is possible to install an integrated in-place washing system to clean the ionizer and collecting plates instead of taking them out for cleaning.

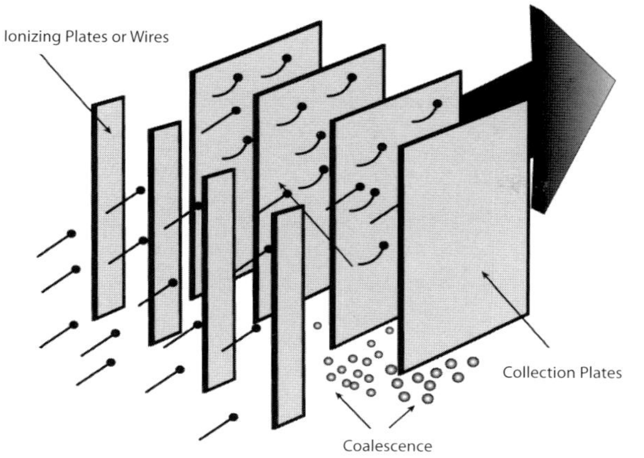

Photo Courtesy of PlymoVent North America

Figure 15.5. Typical Ionizing and Collector Section in an Oil Mist Electrostatic Precipitator

Advantages: Compact design, low pressure drop, and high initial efficiency.

Disadvantages: Frequent cleaning of the collection plates is required to maintain optimum mist and snake removal efficiency. Recurring maintenance can degrade the performance of the cells over time.

Summary

In summary, the nature of oil mist and oil smoke is dramatically different than that of normal airborne contaminant. Basic air filters designed for HVAC systems or for dry dust separators are severely challenged in this environment, which could result in shorter filter service life, and lower than expected filtration efficiency. Therefore, many different varieties of specialized filtration have been developed to remove this contaminant from the airstream.

Through the proper use of machine enclosures, mist collection equipment using the proper levels of filtration, and a comprehensive safety awareness program, the metal working industry can provide a safe and healthy workspace for its employees.

Welding and Other Smoke Removal

Welding, brazing and soldering utilize filler materials that may contain metals that are toxic such as lead and cadmium.

There are more than 80 different types of welding and associated processes. Some of the most common are arc welding, gas-shielded methods of inert gas (MIG), tungsten inert gas (TIG), and plasma arc welding. Other processes may use oxy-acetylene gas, electrical current, lasers, electron beams, friction, ultrasonic sound, chemical reactions, heat from fuel gas and robots.

Welding smoke is a mixture of very fine particles and gases. Many of the substances in welding smoke such as chromium, nickel, arsenic, asbestos, manganese, silica, beryllium, cadmium, nitrogen oxides, phosgene, acrolein, fluorine compounds, carbon monoxide, cobalt, copper, lead, ozone, selenium and zinc can be extremely toxic.

Exposure to welding smoke and fumes such as zinc oxide fumes (welding, cutting or brazing galvanized metals) can cause metal fume fever, with symptoms of chills, fever, thirst, muscle aches and chest pains, coughing, fatigue, nausea and a metallic taste in the mouth. Welding smoke in general may cause a variety of eye, nose, chest and respiratory tract irritation resulting in shortness of breath, bronchitis, pulmonary edema (fluid in the lungs) and pneumonitis (inflammation of the lungs). Long-term chronic health effects may include an increased risk of lung cancer, possible cancer of the larynx and urinary tract.

It is important to capture the contaminants at the source in order that they do not diffuse into the general indoor environment of a factory. Removal of welding smoke and fumes can be

Photo Courtesy PlymoVent North America

Figure 15.6 Source Capture Hood

accomplished using source capture hoods (see Figure 15.6) on media or electrostatic precipitator air cleaners. Some difficult industrial applications such as capturing welding smoke utilize two cells back-to-back in an air cleaner called "double pass." (See Chapter 6.) Recirculation of the air back into the facility requires that it meet strict standards of cleaning. For further information refer to Occupational Health and Safety (OSHA) Regulations 29 CFR, 1910.1026, "Hexavalent Chromium (CrV_1) or contact the American Conference of Governmental Industrial Hygienists (ACGIH). Local jurisdiction or restrictions may also apply. Exhausting to the outdoor air is also an option however, contaminants must not violate local, state or federal air pollution guidelines. It also requires tempering makeup air to the facility in an equal quantity that is exhausted.

Dust Separation

Industrial woodworking, metalworking factories, powder process plants, manufacturing plants and agricultural grain separation are just a few of the applications of dust separation equipment. Although the term "dust" is usually used with this type of contaminant separation i.e. dust separators, larger units can be used to separate any number of types of waste materials.

Dust separation usually involves removal of larger particles of contaminants such as sawdust in woodworking.

Dust separation applications typically have a much higher grain loading than applications discussed up to this point. Dust Separation is used for one of two objectives: 1) recover valuable product from plant processes that generate dust and; 2) clean air to stay in compliance with OSHA and EPA requirements. While many technologies exist for accomplishing these objectives, cyclones separators, baghouse separators, and cartridge separators and are three of the most popular relying on the use of filters to achieve these results.

Dust separation systems are made up of four components:

- Pick Up Points – at the emission source – source capture hoods
- Ductwork – to carry the dust to the collection equipment
- Collection Equipment – i.e., baghouses, cartridge separators, cyclones, ESP's
- Air Mover – such as a fan

Types of Dust Separation Equipment

Cyclone Separators

One of the simplest designs of dust separators is the cyclone separator. With no internal mechanical moving parts other than a blower, the cyclone was designed to clean the air by removing larger particles from fine dusts by inertial separation. See Figure 15.7.

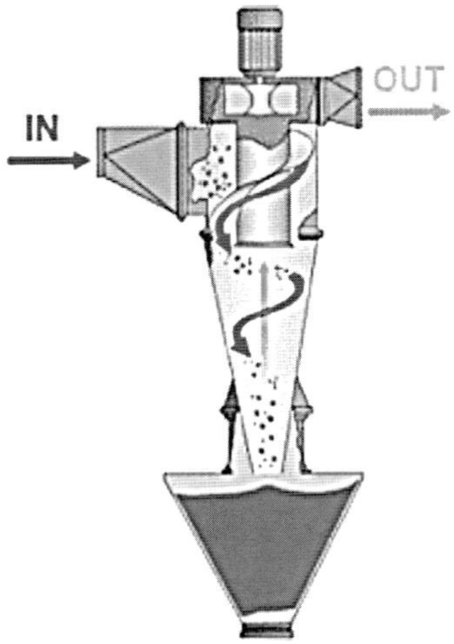

Figure 15.7. Cyclone Separator.

Dust particles are drawn into the separator downward at high velocities into a vortex which causes a cyclone effect, throwing the larger particles against the outside of the cone. Friction with the wall of the separator causes these larger particle to fall down to the bottom and into an air-tight hopper or drum. A tube running up the center of the inside reverses the airflow and the cleaned air moves up this tube to the outlet.

While Cyclone separators can be designed in several configurations depending on the material to be removed, the typical cyclone has

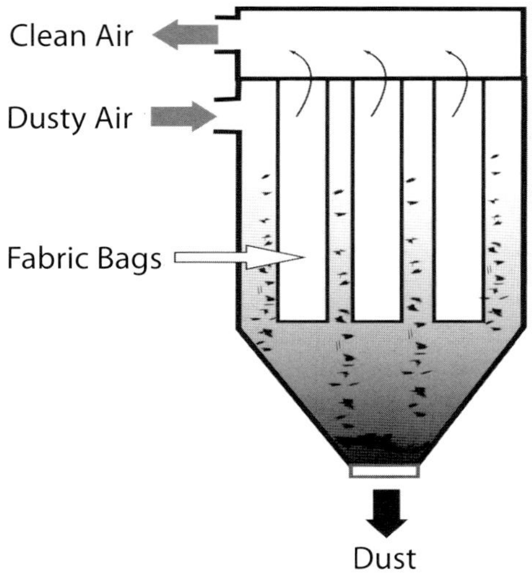

Figure 15.8. Typical Dust Separator Baghouse

a particle removal efficiency of 25 micrometers and larger. Cyclone separators can be fitted with afterfilters, usually of long felt bag material to collect smaller particles, and ducted back into the facility so that makeup air is minimized.

Baghouse Separators

A baghouse is made up of the house, hopper(s), tubesheet, and plenum. Contaminated air enters the house, passes through the bags, or "socks," which are sealed to the tubesheet, and then exits through the clean air plenum. See Figure 15.8. There are three basic baghouse styles: shaker, reverse pulse and pulse jet – each of them relies on a fabric filter to collect the dust and some cleaning mechanism to flex the fabric in order to remove the collected dust from the fabric to the hopper.

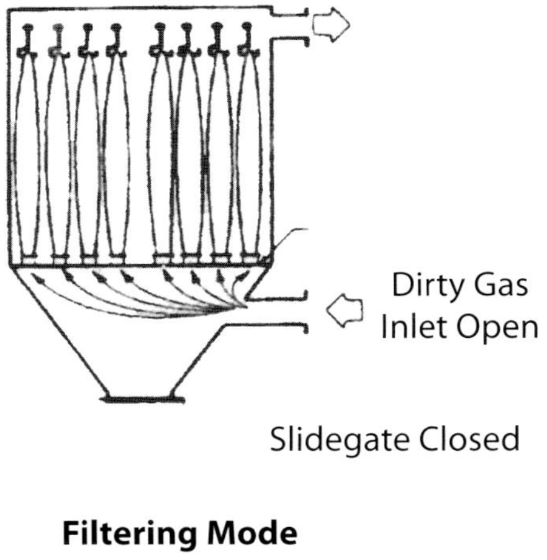

Figure 15.9. Shaker Dust Collector - Cleaning Mode

The oldest technologies, Shaker and Reverse Air baghouses, are similar in that the dust is collected on the inside of the fabric bag or "sock." See Figure 15.9. Shaker baghouses use a mechanical system to gently shake the bags from above. See Figure 15.10.

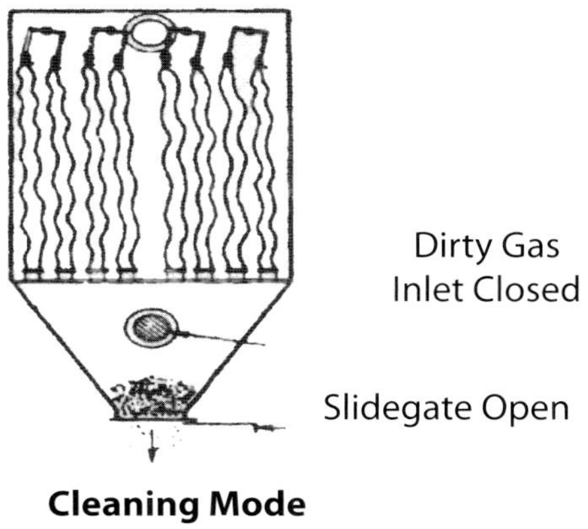

Figure 15.10. Shaker Dust Collector - Cleaning Mode

Reverse Air baghouses are cleaned by shutting off air flow to the baghouse. A blower then moves air in the reverse direction, causing the bags to flex and the dust to fall to the hopper.

The third style of baghouse was introduced in 1956. This Pulse Jet design moved the

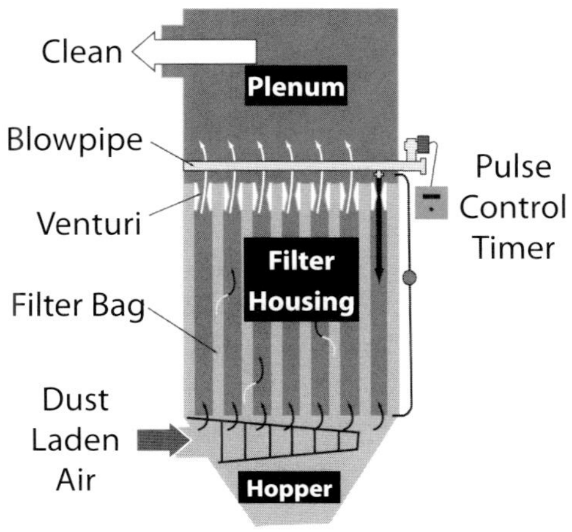

Photo Courtesy Mikropul

Figure 15.11. Jet Pulse Separator

tubesheet to the top of the house so that the dust is collected on the outside of the bag. The bag is cleaned by a burst of compressed air that is directed down the bag from above the tubesheet, In recent years, hybrids have been introduced. See Figure 15.11.

An important consideration in baghouse design is the air to cloth ratio (A:C). This ratio is equivalent to the media velocity in feet per minute. While pulse jet units are operated at 4 to 8 fpm, shaker and reverse air units are operated at 1.5 to 3 fpm. When replacing filters, care should be taken to keep the A:C at or below design conditions.

Cartridge Separators

Cartridge separators are used when space constraints are significant. Cartridge separators are usually pulse jet types and utilize a cartridge filter instead of a bag type filter. Because there are many different types of applications, cartridges are available in many different configurations.

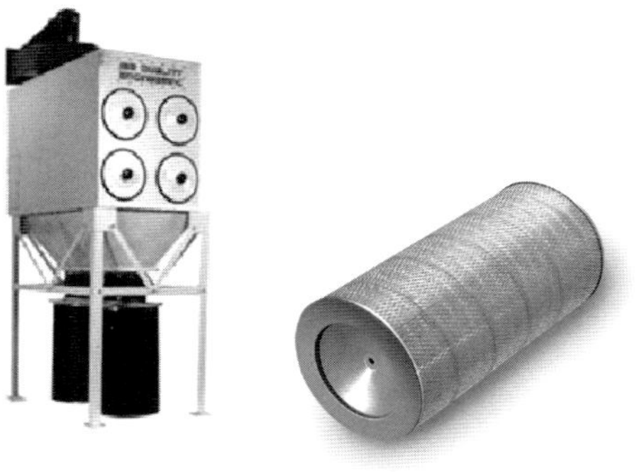

Photo Courtesy Menardi Filters

Figure 15.13. Cartridge Collector and Typical Cartridge

Filter Media Selection

Fabric Media

Fabric selection should consider relative humidity, operating temperature, collection efficiency, particle size and density, air chemistry, dust load, and cleaning method. These variables are factored into the original fabric selection. If a baghouse is having problems or if there have been changes to the process, then other fabrics and filter design should be considered. Many different types of fabrics are available including felt, polyester, acrylic, fiberglass, Teflon, ceramic and more.

Specialized Applications

Industrial collection equipment can be applied to many of the most problematic contaminants. For example, a bag in, bag out housing can be fitted to a dust separator to capture ultrafine dusts that cannot be exhausted such as dust from drug manufacturing plants. Explosive dusts can be captured and contained in downdraft bench applications designed separators that meet NFPA 651-448, NFPA 484, or NFPA 65, An example would be aluminium highway signs that are sanded before painting. Aluminum dust is highly explosive and the velocities for capture along with explosion resistant separators safely and effectively remove this dust. Removal of aviation coatings can involve hazardous materials that can be collected with special dust separation equipment that meets NFPA Std. And high temperature contaminants can be removed and captured in a separator fitted with high temperature bags or cartridges.

Summary

Industrial contaminant control is critical for the health and safety of workers in a facility. Correctly selected and sized for the application air filtration collection equipment can provide a cost effective way to provide solutions to industrial contaminant control.

Notes to Chapter 15

1. “Metalworking Fluids: Safety and Health, Best Practice Manual,” U.S. Department of Labor, Occupational Safety and Health Administration, 1999.
2. National Fire Prevention Assoc. Standard 65, chapters 2, Section 2-1 and 2-4 and NFPA 484 Std. chapter 44.3.1.2.

Appendix One

Systems, Formulas and Conversions

The SI (Standard International) System Measurement

The Standard International (SI) system of measurements is used in most countries throughout the world. However, the Inch-Pound (I-P) system of measurements is used in the United States. Even in the United States, "metric" measurements that are part of the SI system are used extensively for the dimensioning of machinery and fasteners. SI measurements are also used in scientific research. They are the required measurement for all international standards.

In this publication all values are given in the I-P system and are followed, in most instances, by their SI equivalent. This has been done in recognition of the fact that in the United States, the I-P system is currently preferred by most users of air filters. It is the one with which most air filter sales and service technicians are familiar.

There are seven base units in the Standard International System of Units. The ones of interest to the HVAC industry are:

Quantity	Name	Symbol
length	meter	m
mass	kilogram	kg
time	second	s
electric current	ampere	A
thermodynamic temperature	Kelvin	K
amount of substance	mole	mol

Many others are derived from these. A unique characteristic of the SI system is that each physical quantity has only one unit. The base and derived units may be modified by prefixes.

A complete discussion of SI terminology and grammar is too lengthy to be reviewed here. The ASHRAE publication *SI for HVAC and R* is an excellent source for information about the use of SI measurements in heating, ventilating and air conditioning. It not only lists all the SI equivalents for the I-P system, but also explains the use and meaning of various prefixes that are used. It also includes information about the correct grammatical usage of SI terms.

The following SI units are used to describe filter systems:

Measurement	Symbol	Description
Air volume flow rate	m^3/s	Cubic meters per second
Filter pressure drop	Pa	Pascals
Face area	m^2	square meters
Filter dimensions	mm	millimeters
Particle size	μm	Micrometers (ii)
Velocity	m/s	meters/second
Temperature	°C	Degrees Celsius (iii)

Notes:

i. Air volume flow is also given in litres per second (L/s). There are 1000 litres in a cubic meter.

ii. "Micron" is not used in the SI system to give dimensions of very small dust particles. Instead the word "micrometer" is used. The prefix micro means 0.000,001 (10^{-6}).

iii. The temperature in degrees Celsius can be obtained by subtracting 273.15 from degrees Kelvin. At 0°C the thermodynamic temperature is 273.15°K

Useful Formulas:

1. To calculate volume flow (cfm) from velocity (fpm):

$$cfm = fpm \times ft^2$$

2. To calculate velocity (fpm) from volume flow (cfm):

$$fpm = cfm/ft^2$$

3. To calculate velocity (fpm) from velocity pressure (P_v):

$$fpm = 4005 \sqrt{P_v}$$

4. To calculate volume flow (cfm) from velocity pressure (P_v):

$$cfm = (4005 \sqrt{P_v})\ ft^2$$

5. To calculate (a): duct or (b): filter face velocity pressure (P_v) from volume flow (cfm):

$$P_v = [(cfm/ft^2)/4005]^2$$

Symbols for Calculations

A_F	=	filter face area
cfm	=	volume flow of air (cubic feet per minute)
fpm	=	velocity of air (feet per minute.)
P_v	=	velocity pressure (in. w.g.)
ft2	=	(1) cross sectional area of duct (2) face area of filter or filter bank.

Notes

i. SI for HVAC & R, American Society of Heating Refrigerating and Air-Conditioning Engineers Inc., 1791 Tullie Circle, Atlanta, GA 30329

Conversions

Multiply	By	To Obtain
Nanometers	0.0011	Micrometers
Centimeters	0.3937	Inches
Cubic feet	0.28317	Cubic meters
Cubic inches	16.387	Cubic centimeters
Feet	0.3048	Meters
Horsepower	0.7457	Kilowatts
Inches	2.540	Centimeters
Meters	3.281	Feet
Square meters	10.764	Square feet
Cubic meters	35.315	Cubic feet
Meters/second	196.85	fpm
Cubic meters/second	2118.88	cfm
Cubic Meters/hour	0.5886	cfm
Celsius	33.80	Fahrenheit
Pascals	0.00402	Inches w.g.

For further conversions, visit www.onlineconversion.com/auto_convert.pl, or www.sciencemadesimple.net.

Appendix Two

Federal and State Codes and Standards

Codes and Regulations

There are four Federal agencies that have regulations and/or guidelines for the operation of industrial finishing systems:

- The United States Environmental Protection Agency (EPA)
- The Occupational Safety and Health Administration (OSHA)
- National Institute for Occupational Safety and Health (NIOSH)
- American Conference of Governmental Industrial Hygienists (ACGIH)

The EPA is responsible for the environment and as such has authority over what and how much material can be discharged into the atmosphere. The three others are concerned with hazards in the workplace. They set standards on the number of air changes in a finishing room or the velocity of air through the finishing room and through any openings in the room walls, ceilings or floors. They also publish acceptable limits on various VOCs within the workplace.[1] These agencies have their counterparts in every state. Whenever a state regulation is stricter than the Federal regulation, the former prevails. *Information on ventilation rates is best secured from local authorities.*

Impact of the '90s Clean Air Legislation[2]

Title III of the Clean Air Act Amendments of 1990 introduced the concept of regulating emitters based upon proven technology available to individual emitting sources. The EPA was required to publish a list of major source categories emitting one or more hazardous air pollutants (HAPs). By law the source category had to include all categories and subcategories of major sources of the listed HAPs. In July 1992 EPA issued an initial list of 174 industrial source categories. Congress gave the EPA until November 2000 to identify and list the categories of areas responsible for 90% or more of area source emissions and finalize regulations for these sources. As of December 1999 three have been promulgated. They are:

- 40 CFR63 Part IV (Updated 2007) National Emission Standards for Aerospace Manufacturing and Rework Facilities.
- 40 CFR63 (Updated 2007) National Emission Standards for Shipbuilding and Ship Repair.
- 40 CFR63 (1998) National Emission Standards for Wood Furniture Manufacturing Operations.

The aerospace standard (NESHAP)[3] is unique. In section 63.745 paragraph (g) entitled "Inorganic HAP emissions" the EPA introduced its first functional performance test for paint over-spray collection systems, be they wet or dry. Repeated trials demonstrated that it is feasible to mechanically remove very small (less than 10 micron) particulates found in aircraft booth exhaust air streams. The aerospace NESHAP promulgates different efficiencies for "existing" sources and "new source" (sources that commenced construction after October 29, 1996).

Notes to Appendix Two

1. For more information on acceptable limits visit www.cdc.gov/niosh/87104 48/html
2. This paragraph reproduced with permission from the copyright owner, Profit Builders. Information first published in *Technical Advancement in Overspray Collection* (1999) by Rand Schweizer, Chemco Manufacturing Co. Inc. and Jim Motter, Profit Builders
3. NESHAP = National Emission Standard for Hazardous Air Pollutant

Appendix Three

Independent Testing Laboratories Capable of Performing Tests on Air Cleaners (Air Filters) Using ASHRAE Test Methods 52.1 and 52.2

Air Cleaning Technologies Laboratory
University of Illinois at Chicago
842 West Taylor Street
Chicago, IL 60607-7022
www.actlab.uic.edu

Air Filter Testing Laboratories Inc.
4632 Old La Grange Rd.
Crestwood KY 40014
www.aftl.net

Blue Heaven Technologies
2820 S. English Station Road
Louisville, KY 40299
www.blueheaventech.com

Intertek ETL Semko
3933 US Route 11
Cortland, NY 13045
www.intertek.etlsemko.com

LMS Technologies Inc.
6423 Cecilia Circle
Bloomington, MN 55439
www.lmstechnology.com

RTI
Center for Aerosol Technology
Research Triangle Institute
PO Box 12194
Research Triangle Park, NC 27709-2194
www.rti.com

Notes to Appendix Three

1. This listing is for reader's convenience. It is not necessarily a complete listing.
2. The laboratories listed may be capable of performing other tests on air cleaners. They should be contacted directly for any additional information.

Appendix Four

Filter Equilibrium

We often are inclined to think of air filtration as an irreversible action where a dust particle is captured once and stays captured. This may frequently be true of filter media incorporating adhesive-coated fibers, but is not true of dry-media filters, especially those in the high-efficiency range.

A particle may be captured by a fiber only to become reentrained in the airstream and then be captured b another fiber downstream of the first one. This cycle may be repeated. Eventually the particle either becomes strongly adhered to a fiber or passes through the filter.

For a dust particle to become permanently captured by a fiber, it is necessary that the forces of adhesion between the fiber and the particle be equal to or greater than the drag force of the air as it moves past the fiber-particle zone.

Dutch[1] reported that this could be expressed mathematically by the formula:

$$F_A \geq F_S$$

Where:
F_A = Force of adhesion
F_S = Drag force of air on a dust particle

$$F_A = k\,(r_1 + r_2) / r_1 r_2$$

Where:
k = 40 (a constant)
r_1 = Particle radius
r_2 = Fiber radius

$$FS = 3.4\ rv \times 10^{-3}\ \text{dynes}$$

Where:
r = Particle radius
v = Media velocity

Dutch reported that, using this formula, he was able to get acceptable correlation between calculated data and filter media efficiency measurements.

Examination of the formula for the forces of adhesion shows that it will be the strongest when both fiber and particle are the same size and will become stronger as fiber and particle become smaller.

The shear force, on the other hand, will become stronger as the particle size and/or the media velocity increases.

Notes to Appendix Four

1. Source: Notes made of a presentation by Paul Dutch at a Contamination Control Conference sponsored by the AACC. Pasadena CA, 1965

Appendix Five

A Brief Description of the ANSI/ASHRAE Standard 52.1 Test Method

To fully understand the results of a filter test, one should be familiar with the general setup of the test equipment, the testing procedure and the report form showing the results of the test.

Test Equipment

All the basic components of the ANSI/ASHRAE Standard 52.1 test duct are shown in Figure Five-1. These various components are identified by number. Readers interested in more details about any of these components should refer to the original document.[1]

The Test Procedure

1. The same setup is used for both efficiency and arrestance testing except that a removable section including the final filter is used for arrestance but is replaced by a filler duct section when the efficiency test is being run.
2. There are two alternative duct arrangements, one with the blower upstream of the sampling setup and the other with the blower downstream of all of the sampling equipment. The downstream version is shown in Figure Five-1.
3. Sampling probes for efficiency tests are described in detail and are designed so that both the upstream and downstream probes sample at the same rate. This is done by the use of flow-limiting critical flow venturi nozzles in the sampling line. Once the nozzles are matched, there is no need to measure the sampling airflow.
4. The target paper used for dust spot testing is an all-glass HEPA filter media.

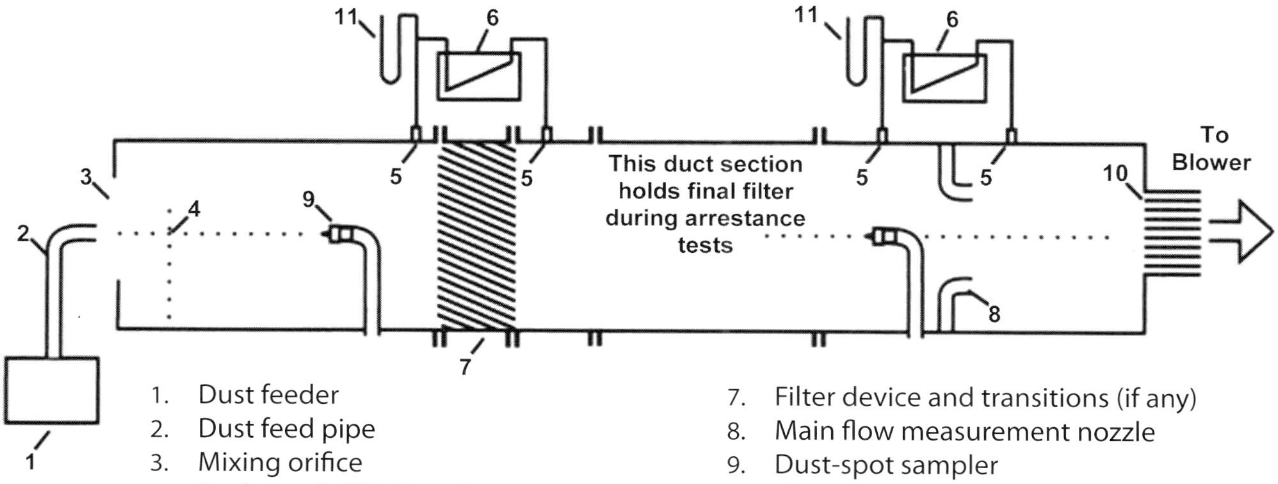

1. Dust feeder
2. Dust feed pipe
3. Mixing orifice
4. Perforated diffusion plate
5. Static tap
6. Manometer
7. Filter device and transitions (if any)
8. Main flow measurement nozzle
9. Dust-spot sampler
10. Flow straightener
11. Vertical manometer for measurement of nozzle inlet pressure

Figure Five-1. ASHRAE Air Filter Test Duct, Blower Downstream of Test Filter

5. The synthetic dust feeder is detailed and is designed to feed dust at the rate of 2 0.2 g/1000 ft^3 (71 7 mg/m^3) of test air.
6. Criteria are given for the final filter to be used in arrestance testing.
7. The opacity meter and methods for calibration are fully described. Using this equipment, the relative light transmission characteristics of the target papers before and after filter tests can be determined. From this data the opacity of the target can be calculated.

Efficiency

Efficiency is determined by comparing the discoloration of two targets, one sampling atmospheric air upstream of the filter under test and the other sampling on the downstream side of the filter. Since the filter will remove some of the staining portion of the atmospheric dust, the downstream target (after the filter) will discolor more slowly than the upstream one.

Filters with average efficiencies below 20% or above 98% are not considered suitable for this test method. Filters with low efficiencies can be reported as "less than 20%." Filters with efficiencies above 98% are typically tested by the-DOP test procedure or other methods used for HEPA filters. (See Chapter 8: *"HEPA and ULPA Filter Testing."*)

There are two different allowable methods for measuring efficiency: 1) The intermittent flow method; and 2) The constant flow method.

Intermittent Flow Method

In the intermittent flow method the downstream probe is sampling continuously. A sampling time is selected for the upstream probe which will cause its target to discolor at the same rate as the downstream probe. For instance, if a filter estimated to be in the 90% range is being tested, the upstream probe should be used 6 seconds out of every 60 seconds (that is, 10% of the total time). The downstream probe is operated continuously during the test and the upstream probe sampling time is controlled by a 1-minute recycling timer connected to a solenoid valve in the sampler line. In our example below, the timer opens the valve for 6 seconds out of every minute (10% of the total time).

It is unlikely that the two targets will darken at exactly the same rate. To compensate for this, the opacity (reduction in light transmission of the target) is measured for both the upstream and downstream target papers and this is factored into the equation for determining efficiency:

$$E = 100 \times [1 - (S_u/S_d)(Y_d/Y_u)]$$

Where:

E = Dust spot efficiency

S_u = Total "on" time for the upstream sampler during the test

S_d = Total "on" time for the downstream sampler during the test

Y_u = Opacity of the upstream target

Y_d = Opacity of the downstream target

As an example, if the filter was exactly 90% efficient and we sampled the upstream target for 6 seconds out of every minute, the opacity of the targets would be the same (let's assume it would be 30%) and the equation would become:

$$E = 100 \times [1 - (6/60) \times (30/30)]$$
$$= 100 \times [1 - (.1 \times 1)] = 90\%$$

If the opacities were not the same (let's assume the upstream was 30% and the downstream was 32%), the equation would become:

$$E = 100 \times [1 - (6/60) \times (32/30)]$$
$$= 100 \times [1 - (.1 \times 1.07)] = 89.3\%$$

The Standard requires that this test be repeated and that the results of the two tests must be within defined ranges to be acceptable. If they are, the two values are averaged, and this is reported as the efficiency for that test point. If they are not, three additional tests must be run and the *median* of all five tests is reported as the efficiency.

Since the word "median" is not encountered very often, its meaning should he clarified. Suppose that the results of 5 filter tests which had to be performed were 80.0%, 82.5%, 84.0%, 85.5%, and 86%. The average would be the sum of all the test results (418) divided by 5 or 83.6%. The mean value would be half way between the highest and lowest value or 83%. The median is the value which has an equal number of results higher and lower than it. In this case it is 84%.

Constant Flow Method

In this method, both samplers are run for the same period of time. The opacities are calculated as in the intermittent flow method but are not used in the equation because the relationship between opacity and discoloration is not linear. Opacity is used to determine the opacity index which is used instead. The opacity index can be found by reference to an equation given in the Standard, or from a graph plotting opacity against opacity index. These are included in the Standard. The equation for efficiency becomes:

$$E = 100 \times [1 - (Z_d/Z_u)]$$

Where:

Z_d = Opacity index corresponding to the opacity of the downstream target
Z_u = Opacity index corresponding to the opacity of the upstream target

Example: If the upstream opacity is 32.5%, the upstream opacity index read from the curve is 50. Similarly, if the downstream opacity is 4%, its opacity index read from the curve is 5. The equation given above becomes.

$$E = 100 \times [1 - (5/50)] = 100 \times [1 - 0.1]$$
$$= 100 \times .9 = 90\%$$

Two tests must be run. If the two are within a specified allowable range, the two values are averaged and the average reported as the efficiency. If allowable agreement is not achieved, three more tests must be run and the median of all five values is reported as the efficiency.

This method provides a measure of efficiency much more quickly than the intermittent flow method. The sample must be run continuously until the upstream sample obtains an opacity between 20% and 40%. The results of both methods have been found to be equivalent.

Arrestance

In the ASHRAE synthetic dust weight arrestance test, the nature of the synthetic dust is clearly defined. It is:

a. 72% Standardized Air Cleaner Test Dust, Fine.

This is classified from dust gathered in a desert area of Arizona. It is predominantly silica with a mass-mean diameter of 7.7 μm. The weight percentage of particles of different size contributed by this material to ASHRAE Test Dust is:

Particle Size	Percent
0-5	28.0%
5-10	13.0
10-20	11.5
20-40	13.0
40-80	6.5

b. 23% Powdered carbon (carbon black in powdered form meeting listed specifications).
c. 5% Cotton linters #7 ground in a Wiley Mill with a 4mm screen.

All percentages are by weight.

The test procedure requires that any dust in excess of 1% of the amount fed that settles in the upstream test duct section be reentrained so that all the dust fed is either on the filter under test or in the final filter downstream of the unit under test. It defines arrestance for any particular test period as:

$$A = 100\,[1 - W_d/W_u]$$

Where:
A = Arrestance, percent
W_u = Weight of synthetic dust fed
W_d = Weight of dust captured on final filter

Testing Sequence

The sequence of testing is as follows:

1. Measure the pressure drop of the clean filter under test at 50, 75, 100, and 125% of rated airflow.
2. Run the first atmospheric dust spot efficiency test (on clean filter).
3. With the final filter duct section in place, load the mechanical dust feeder with a weighed amount of ASHRAE test dust. Ideally this should be 25% of the total expected to be fed. Feed the dust at a rate of 2 grams per 1000 cubic feet (71 mg/m^3) of test air. At the end of the segment record the amount of dust fed and measure the amount of dust captured on the final filter.
4. Repeat the atmospheric dust spot efficiency test.
5. Repeat step #3, loading second segment of dust.
6. Repeat steps #4 and #5 until one of the following occurs:
   a. The rated final pressure drop is reached or exceeded by a whole dust increment.
   b. Two consecutive arrestance measurements are less than 85% of the maximum arrestance measured during the run.
   c. The arrestance is less than 75% of the maximum arrestance measured, per cycle, for self-renewing filters.
7. At the end of the test, calculate the average ASHRAE dust spot efficiency, average ASHRAE arrestance, and dust holding capacity using the procedure outlined in the Standard.

Self-Renewable Devices

It should be noted that there are special procedures in the Standard for the testing of various types of self-renewing filters such as automatic roll filters. These procedures are complex and are not included here.

Filter Test Report Form

A standard report form describing a filter test can provide a great deal of information about that filter to a buyer or seller. While testing organizations are not required to adopt the exact report form suggested by the Standard, it requires that all the items shown in the suggested version be included in any version that they may use. The summary consists of two pages. The various information keyed on page 1 and shown in Figure Five-2 is as follows:

1. ***Name of Laboratory Performing the Test.*** This laboratory should be acceptable to the end user.
2. ***Test Method Used.*** This would be 52.1. 52-76 is now obsolete.
3. ***Report Number.*** This is important for identification. Independent laboratories will reference a report number to clarify any test data. If the report covers several tests, this section will identify which one is being reported.
4. ***Originator of Test.*** This shows who requested the test, who made the filter, and what the name of the product was.
5. ***Source of Filter.*** This shows whether the filter was supplied by the manufacturer or purchased on the open market.
6. ***Filter Model Number and Dimensions.*** This is important for identification.
7. ***Published Information about the Filter.*** This is useful information comparing published information with actual test results. It is not usually required by most specifications.
8. ***Test Results.*** This is the most important part of the report and includes the following data:
   a. ***Test Airflow Rate.*** This states the airflow rate (cfm or m^3/s) through the filter during the test. Note that this can be the manufacturer's rated flow or some other value.

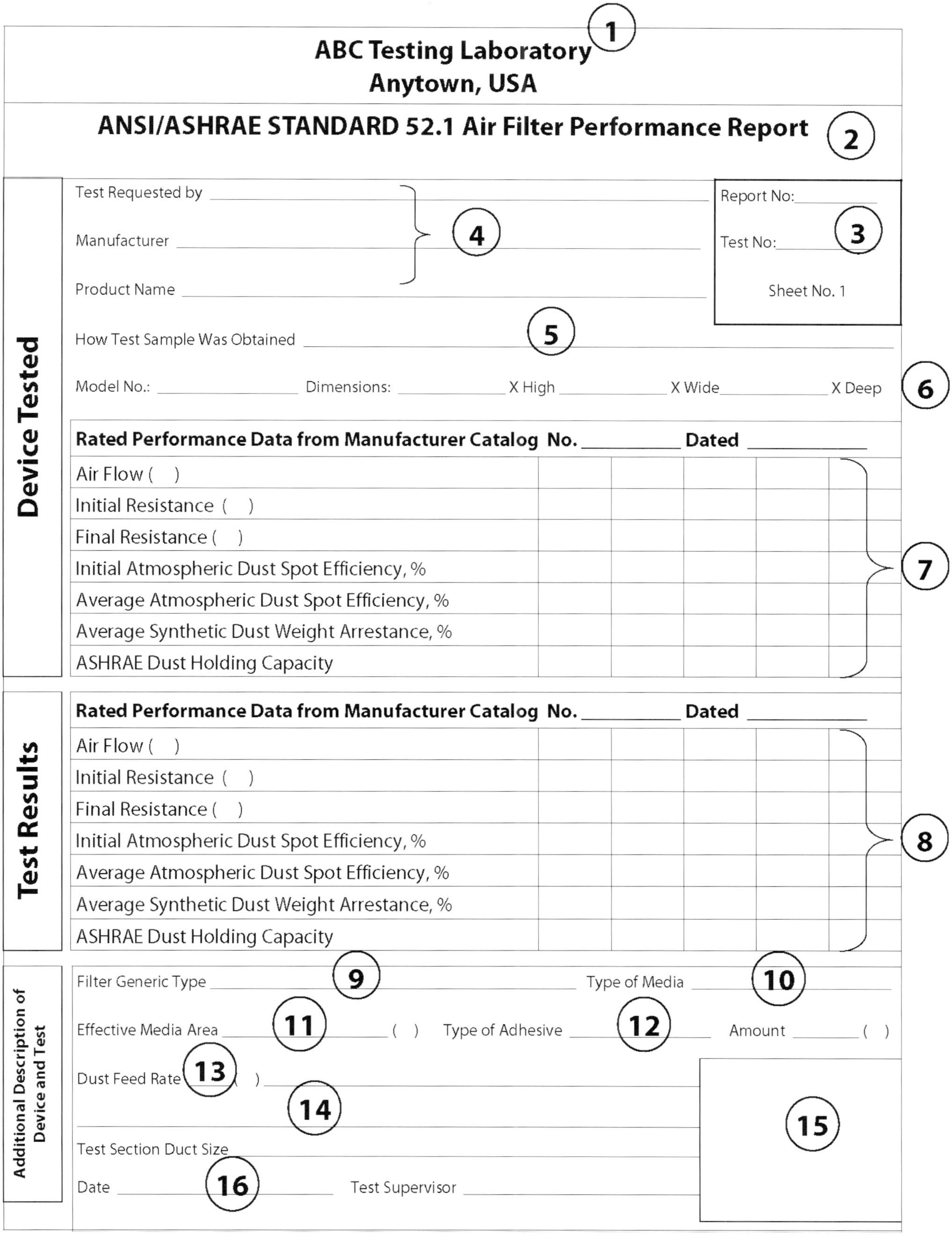

Figure Five-2. Suggested ASHRAE Air Filter Performance Report Form

b. ***Initial Resistance.*** This is the resistance of the clean filter to the airflow. It is expressed either in inches of water (in. w.g.) or Pascals (Pa).

c. ***Final Resistance.*** This is the resistance at which the test was terminated. In comparing the ASHRAE atmospheric dust spot efficiency of two filters, it is important that their test reports show that they were both run to the same final pressure drop. In comparing the ASHRAE arrestance of two filters, one may not reach the same final pressure drop as the other because its arrestance falls below the values specified in the ASHRAE test method.

d. ***Initial Atmospheric Dust Spot Efficiency***. This is the ASHRAE atmospheric dust-spot efficiency of the filter when it is clean. Since efficiency usually increases with test dust loading, this may be the minimum Efficiency of the filter.

e. ***Average Atmospheric Dust Spot Efficiency.*** It is the intent of the ASHRAE test method that, wherever possible, the procedure include four increments of synthetic dust feeding. Average ASHRAE atmospheric dust spot efficiency is not the mean of the initial and final reading. Instead, a procedure is specified which takes into account the amount of dust fed between increments and the ASHRAE atmospheric dust spot efficiency measured after each reading.

f. ***Average Synthetic Dust Weight Arrestance.*** The arrestance is measured in increments, and not merely by feeding the synthetic dust at one time until a desired final pressure drop is reached. The ASHRAE test procedure defines a method for using the arrestance measured at the end of each increment and determines the Average Arrestance.

g. ***ASHRAE Dust Holding Capacity.*** This is a sum of individual dust holding capacities run in the test segments. For each segment, dust holding capacity is the weight of the dust fed less the weight of the dust passing the filter under test. The dust passing the test filter is measured by the weight increase of the final filter.

9. ***Filter Generic Type.*** This assists in the description of the filter.

10. ***Type of Media.*** This also assists in the description of the filter.

11. ***Effective Media Area.*** This is a measure of the usable media as contrasted with the total amount that may be in the filter.

12. ***Type of Adhesive.*** The amount and type of adhesive may affect the performance of a panel filter. The report gives this information.

13. ***Dust Feed Rate.*** This is the rate which dust is introduced into the test fixture (typically 2.0 grams per 1000 cubic feet of air).

14. ***General Remarks.*** This is a brief summary of data that may not have a specific space on the report. For instance, it may describe the header construction in a non-supported extended surface filter.

15. ***Seal.*** This is space for the seal of a professional engineer who may have performed the test or supervised the testing.

16. ***Date.*** This does not specifically state whether this is the date the test was performed or whether it is the date the report is written. Usually it will be the same. It does serve the useful purpose of confirming the age of the report. Some specifications require that test reports on filters be less than 3 years old.

Supplementary Data

Page 2 of the suggested Summary Report Form displays graphically the results of the tests that were performed. Three charts give valuable information about filter performance throughout the test. These are:

1. The ASHRAE atmospheric dust spot efficiency and synthetic dust weight arrestance of the filter at different points during the loading of the filter with synthetic dust.
2. The increase in pressure drop of the filter as dust is fed to it. Pressure drop is reported in inches of water or Pascals.
3. The resistance of a clean filter at different airflows through it. Resistance is shown in inches of water or Pascals.

In addition to these summary sheets there are also worksheets in which test readings are noted and calculations made.

Note to Appendix Five

1. ANSI/ASHRAE Standard 52.1 - 1992 Gravimetric and Dust Spot Procedures for Testing Air Cleaning Devices Used in General Ventilation for Removing Particulate Matter, American Society of Heating, Refrigerating and Air-Conditioning Engineers Inc., 1791 Tullie Circle NE, Atlanta GA 30329.

Appendix Six

Methods of Testing Overspray Filters

Filter Testing Methods

There are three identifiable methods that have been used for testing air filters to determine their efficiency in removing overspray. They are compared in Figure Six-5.

AFTL Method

The AFTL proprietary method developed by Air Filter Testing Laboratories at the request of individual filter manufacturers uses a modified ASHRAE 52.1 test duct. (See Chapter 7: *"HVAC Filter Testing"*). The preweighed filter to be tested is installed in the same relative location as it would be for a 52.1 synthetic dust arrestance test. Likewise a preweighed (arrestance) final filter is located downstream of the test filter in the same relative location as in 52.1. Paint is sprayed into the operating test system until the filter under test reaches a designated final pressure drop. At the completion of the test, both the filter under test and the final filter are reweighed. The percent efficiency of the filter can be calculated by the formula:

$$E = (I_T / [I_T \; I_F]) \times 100$$

Where:

E Paint removal efficiency

I_F weight increase of final filter

I_T weight increase of test filter

Test results include overspray removal efficiency, paint holding capacity, and initial and final pressure drops at a stated filter face velocity.

LMS Technologies

LMS Technologies Inc.[1] has developed a proprietary method of testing filters for their ability to remove overspray by modifying an ASHRAE 52.2 test duct and introducing a spray gun to spray paint against a target, both located within the test duct. The filter under test is located in the same relative position as in 52.2. (See Chapter 7: *"HVAC Filter Testing"*) While the test is in progress, air is sampled upstream and downstream of the filter under test. Portions of the sampled air are measured by aerodynamic particle sizers that give the particle count and size distribution of the samples upstream and downstream of the filter. Using this information, the fractional efficiency of a filter on particles from 0.3 to 10 microns can be reported. The procedure calls for the test rig to be run with the spray in operation but without the test filter in place. When measurements indicate that the spray is stabilized, the test filter is installed and 8 runs are made of particle size and distribution. The sampling is discontinued, but the paint spray may be kept on for some additional time. 7 minutes seem to be the most frequently used. After this period, samples are taken again with the paint spray still running. A comparison between the runs shows the effect of filter loading on efficiency. In this test the pressure drop of the clean filter under test is measured.

EPA Method 319

This test[2] uses a duct identical to the one used in ASHRAE 52.2 except that there may be slight differences in the aerosol generating section. As in ASHRAE 52.2 one of the challenges is a potassium chloride solid aerosol formed by the evaporation of a concentrated solution of potassium chloride. However, in addition to the potassium chloride aerosol, oleic acid, a liquid, is sprayed in such a way as to form small particles. The potassium chloride is intended to simulate dry paint overspray and the oleic acid, wet overspray.

The whole test program is performed essentially as described in ASHRAE method 52.2. The two challenges are fed separately. Separate measurements and calculations are performed to generate fractional efficiency curves showing the ability of the filter to remove particles of each challenge. Fractional efficiency is given for specific sizes in the range from 0.3 to 10.0 microns. The clean filter (paint arrestor) pressure drop is measured and recorded. The fractional efficiency is calculated and recorded.

This method is unique in the extent to which it requires checks and rechecks to make certain that the results are accurate. In that respect it is again similar to, if not a duplicate of the ASHRAE 52.2 method.

Application of Method 319

The aerospace standard (NESHAP) is the first to specify Method 319.

For existing sources, (construction or reconstruction commenced before October 29, 1996) the owner or operator is given several choices as to methods of control. The one of interest to us states:

Before exhausting it to the atmosphere, pass the air stream through a dry particulate filter system certified using the methods described in 63.750(o) to meet or exceed the efficiency data points in Figures Six-1 and Six-2 . . .

Figure Six-1. Two-Stage Arrestor, Liquid Phase Challenge for Existing Sources

Filtration Requirements %	Aerodynamic Particle Size range mm
>90	> 5.7
> 50	> 4.1
>10	> 2.2

Figure Six-2. Two-Stage Arrestor, Solid Phase Challenge for Existing Source

Filtration Requirements %	Aerodynamic Particle Size range µm
> 90	> 8.1
> 50	>5.0
>10	>2.6

Figure Six-3. Three-Stage Arrestor; Liquid Phase Challenge for New Sources

Filtration Requirements %	Aerodynamic Particle Size range µm
>95	> 2.0
>80	> 1.0
>65	> 0.42

Figure Six-4. Three Stage Arrestor; Solid Phase Challenge for New Sources

Filtration Requirements %	Aerodynamic Particle Size range µm
>95	> 2.5
> 85	> 1.1
> 75	> 0.70

A two-stage filter system is described as: "dry particulate filter system using two layers of filter media to remove particulate. The first stage is designed to remove the bulk of the particulate and a higher efficiency second stage is designed to remove smaller particulates."

For new sources using control by a dry particulate filter, the system must meet or exceed the efficiency data points in Figures Six-3 and Six-4.

A three stage arrestor is not defined. Instead recent EPA rulings have clarified that the "two stage" and "three stage" designations define a functional level of performance and are not to be understood as literal requirements for 2 or 3 stages of filtration.

Regardless of the number of stages, final efficiency is based on measuring the concentration of test aerosol downstream of the entire filter system used as an assembly. The efficiencies of each filter are not as important as the efficiency of the sum of all the parts.

Anticipated Method 319 Impact[3]

The implementation of the aerospace NESHAP raises the "bar" for paint overspray collector technology beyond aerospace finishing. In the foreseeable future, similar regulations will be applied to other finishing sources utilizing coatings containing HAPs. EPA is required to apply technologies developed in one industry sector to other industry sectors with the similar HAP emissions. During 1999, unofficial EPA sources indicated that the agency is considering amending the wood furniture NESHAP to require overspray collectors meet performance standards similar to those presently promulgated for the aerospace industry. Furthermore, similar changes are being considered for pending standards applicable to metal and plastic finishing operations.

The availability of collector FP test data (capture efficiency as a function of droplet size, using paint as a challenge) further raises the maximum control technology. It will precipitate additional changes in EPA's regulation of finishing operations. In short, the aerospace NESHAP will initiate an avalanche of changes in both wet and dry paint overspray control technology. In the coming decade these changes will significantly impact overspray collector design.

Notes to Appendix Six

1. K.C. Kwock, Filtration Characteristics of Paint Booth Filters, Air Media, Summer 1999 (pp 8-13)
2. Appendix A to 40CFR 63.745 (g)—Test Methods: Method 319: DETERMINATION OF FILTRATION EFFICIENCY FOR PAINT OVERSPRAY ARRESTORS
3. Paragraph reproduced with permission by the copyright owner, Profit Builders. (See Note 2 Appendix Two.)

Method	Duct Section (1)	Challenge	Paint Baffle	Filter Loading	Report Items
AFTL	ASHRAE 52.1	Paint Spray	No	Yes to desired final pressure drop	Initial Pressure Drop Final Pressure Drop Filter Efficiency Paint Holding Capacity
LMS Technologies	ASHRAE 52.2	Paint Spray	Yes	Yes prior to second test	Initial Pressure Drop Fractional Efficiency
EPA Method 319	ASHRAE 52.2	Potassium chloride Oleic acid	NA	No	Initial Pressure Drop Fractional Efficiency

Note:
1. Modified to introduce proprietary features.

Figure Six-5. Comparison of Paint Arrestor Test Procedures.

Appendix Seven

Contaminant Capacity Index

*Acetaldehyde	2
Acetic acid	4
Acetic anhydride	4
Acetone	3
*Acetylene	1
*Acrolein	3
Acrylic acid	4
Acrylonitrile	4
Adhesives	4
Air-Wick	4
*Amines	2
*Ammonia	2
Amyl acetate	4
Amyl alcohol	4
Amyl ether	4
Anesthesics	3
Aniline	4
Automobile exhaust	3
Borane	3
Bromine	4
Butadiene	3
Butane	2
Butonone	4
Butyl acetate	4
Butyl alcohol	4
Butyl cellosolve	4
Butyl chloride	4
Butyl ether	4
*Butylene	2
*Butyne	2
*Butyraldehyde	3
Butyric acid	4
Camphor	4

Caprylic acid	4
Carbolic acid	4
Carbon disulfide	4
*Carbon dioxide	1
Carbon monoxide	1
Carbon tetrachloride	4
Cellosolve	4
Cellosolve acetate	4
Chlorine	3
Chlorobenzene	4
Chlorobutadiene	4
Chloroform	4
Chloronitropropane	4
Chloropicrin	4
Corrosive gases	3
Cresote	4
Cresol	4
Crotonaldehyde	4
Cyclohexane	4
Cyclohexanol	4
Cyclohexanone	4
Cyclohexene	4
Decane	4
Dibromethane	4
Dichlorobenzene	4
Dichlorodifluomethane	4
Dichloroethane	4
Dichloroethylene	4
Dichloroethyl ether	4
Dichloromonofluormethane	3
Dichloronitroethane	4
Dichloropropane	4
Dichlorotetrfluoroethane	4

*Diethylamine	3
Diethyl ketone	4
Dimethylaniline	4
Dimethysulfate	4
Dioxane	4
Dipropylketone	4
Ethane	1
Ether	3
Ethyl acetate	4
Ethyl acrylate	4
Ethyl alcohol	4
*Ethyl amine	3
Ethyl benzene	4
Ethyl bromide	4
Ethyl chloride	3
Ethyl ether	3
Ethyl formate	3
Ethyl mercaptan	3
Ethyl silicate	4
*Ethylene	1
Ethylene chlorhydrin	4
Ethylene dichloride	4
Ethylene oxide	3
Fluorotrichloromethane	3
Formaldehyde	2
*Formic acid	3
Heptane	4
Heptylene	4
Hexane	3
Hexylen	3
*Hexyne	3
Hydrogen	1
*Hydrogen bromide	2
*Hydrogen chloride	2
*Hydrogen cyanide	2
*Hydrogen fluoride	2
*Hydrogen iodide	3
*Hydrogen selenide	2
*Hydrogen sulfide	3
Indole	4
Iodine	4
Isophorone	4

Irritants	4
*Isoprene	3
Isopropyl acetate	4
Isopropyl alcohol	4
Isopropyl ether	4
Kerosene	4
Lactic acid	4
Lysol	4
Mercaptans	4
Mesityl oxide	4
Methane	1
Methyl acetate	3
Methyk acylate	4
Methyl alcohol	3
Methyl bromide	4
Methyl buty ketone	4
Methyl cellosolve	4
Methyl cellosolve acetate	4
Methyl chloride	3
Methyl Chloroform	3
Methyl ether	3
Methyl ethyl ketone	4
Methyl formate	3
Methyl isobutyl ketone	4
Methyl mercaptan	4
Methylcyclohexane	4
Methylcyclohexanone	4
Methylcyclohexanol	4
Methyl chloride	4
Monochlorobenzene	4
Monofluorotrichloromethane	4
Naptha (coal tar)	4
Naptha (petroleum)	4
Naphthalene	4
*Nitric acid	3
*Nitro benzenes	4
Nitroethane	4
*Nitrogen dioxide	2
Nytroglycerine	4
Nitromethane	4
Nitropropane	4
Nonane	4

Octalene	4
Octane	4
Ozone	4
Palmitic acid	4
Paradichlorobenzene	4
Pentane	3
Pentanone	4
*Pentylene	3
*Pentyne	3
Perchloroethylene	4
Phenol	4
Phosgene	3
Propane	3
Propionaldehyde	3
Propionic acid	4
Propyl acetate	4
Propyl alcohol	4
Propyl chloride	4
Propyl ether	4
Propyl mercaptan	4

*Propylene	3
*Propyne	2
Pyridine	4
Styrene monomer	4
*Sulfur dioxide	3
*Sulfur trioxide	3
Sulfuric acid	4
Tear gas	4
Tetrachloroethane	4
Tetrachloroethylene	4
Toluene	4
Toluidine	4
Trichlorethylene	4
Trichloroethane	4
Turpentine	4
Uric acid	4
Valeric acid	4
Valericaldehyde	4
Vinyl chloride	3
Xylene	4

Some of the contaminants listed in the table are specific chemical compounds. Some represent classes of compounds and others are mixtures and of variable composition. Activated charcoal capacity for odors varies somewhat with the concentration in air with humidity and temperature. The numbers given represent typical or average conditions and might vary in specific instances.

The capacity index has the following meaning:

4. High capacity for all materials in this category. One pound takes up about 20% to 50% of tis own weight - average about 1/3 (33-1/3%). This category includes most of the odor causing substances.

3. Satisfactory capacity for all items in this category. These constitute good applications but the capacity is not as high as for category 4. Absorbs about 10 to 25% of its weight-average about 1/6 (16.7%).

2. Includes substances which are not highly adsorbed by which might be taken up sufficiently to give good service under the particular conditions of operation. These require individual checking.

1. Adsorption capacity is low for these materials. Activated charcoal cannot be satisfactorily used to remove them under ordinary circumstances.

**For the asterisked compounds, treated impregnated charcoal may be used to remove trace compounds from the air.*

Appendix Eight

Service Life of Molecular Air Filtration Media

After a molecular air filtration system has been purchased and installed, what one ultimately wants to know is, "What is the expected service life of the filter media?" Service life can be defined in many ways, but is most often referred to in terms of removal efficiency (i.e., when the efficiency of the system drops below a minimum value, the molecular filter media is replaced). What this minimum efficiency may be is the subject of much debate. It may be an actual value (e.g., 50%) or can be based on a subjective criteria such as odor. Regardless of how service life is defined, it would be of considerable benefit if a relatively simple expression could be used to predict service life.

The service life is directly proportional to removal capacity and weight of the adsorbent. It is inversely proportional to the adsorption efficiency and airflow of the system, and molecular weight and concentration of the target contaminant. This can be represented by the following formula.[1,2,3]

Equation I

$$t = (6.43 \times 10^6) SW / EQMC$$

Where:

t = Duration of the adsorbent service before saturation (hours)

S = Adsorptive capacity of the media (fractional)

W = Weight of adsorbent (pounds)

E = Removal efficiency (fractional)

Q = Airflow rate through the adsorbent bed (cfm)

M = Average molecular weight of the adsorbed vapor (gm/mole)

C = Inlet vapor concentration (ppm by volume)

Note that Equation 1 implies a total saturation of the adsorbent, where the removal efficiency is essentially zero. It is unlikely, that one would allow a gas-phase system to reach saturation before the media is changed; therefore, one might pick a target efficiency and factor the time accordingly.

Another model has been developed to estimate the service time to a 10% breakthrough (90% removal efficiency) and is represented by the following formula.[4]

Equation 2

$$^{t}10\% = (2.4 \times 10^6) w_c (a + bt) / C_{2/3} MQ$$

Where:

$^{t}10\%$ = Service time of the adsorbent to a 10% breakthrough (90% efficiency, minutes)

w_c = Weight of adsorbent (grams)

a,b = Calculated coefficients

t = Boiling point of the vapor (°C)

C = Is the inlet vapor concentration (ppm by volume)

M = Molecular weight

Q = Airflow rate (litres/min)

Compared to Equation 1, Equation 2 has several disadvantages: (a) Only the time to a 10% breakthrough can be calculated; (b) It cannot adequately deal with solvent mixtures; and (c) No provisions are made for differences in chemical properties of the adsorbates. It cannot be used as a predictor for acid gases such as sulfur dioxide, but is applicable for various volatile organic compounds (VOC).

Of these two mathematical models for estimating service life on a relative basis, Equation 1 is more meaningful to IAQ applications. However, its proper use indicates obtaining realistic values for adsorption efficiencies and capacities. Laboratory and/or field testing may be indicated to provide the data necessary to make reasonable service life projections and product comparisons.

Whereas these mathematical models can be used for estimations of media service life and to compare competitive products, they are really not much help once the system is installed and operating. As mentioned earlier, the media life in a gas-phase filter system depends on many variables. In the absence of proper system maintenance, the first indication that the media may be spent and needs to be replaced may come from building occupants themselves in the form of complaints pertaining to odors or irritation. Because of the subjective nature of many IAQ problems, this may be the most reliable indicator of media life. However, one would hope that through preventative maintenance, the media can be replaced on a timely basis in order to head off IAQ-related complaints.

Some media manufacturers offer media analysis services for the expressed purpose of assisting their customers in determining the proper media changeout intervals. A sample of the media is collected from the system and sent to the manufacturer for analysis. The analysis typically involves a determination of the activity remaining in the sample, comparing that to new media, and using the time in service to estimate how much longer the media may be effective. This technique has proven to be particularly effective for those chemically impregnated medias due to the relatively simple analysis involved in determining the amount of unreacted impregnant remaining on the media.

Plain-activated carbons cannot be reliably tested for media life due to problems associated with off-gassing of adsorbed contaminants. One method which is showing some promise is to weigh a carbon sample, heat it to drive off the adsorbed gases, weigh it, heat it once more to drive off any adsorbed water, and weigh it a third time. By comparing its "dry" weight to the weight of the original sample one can determine how much has been adsorbed. Knowing the total adsorptive capacity of the carbon, one can then estimate the relative saturation of the carbon. However, this method only calculates the total weight gain of the carbon due to adsorbed gases and cannot differentiate between those contaminants which have a high or low adsorptive capacity.

The most reliable, accurate, and certainly the most expensive way to gauge media life is by the use of gas-monitoring equipment. If critical control of a particular contaminant or group of contaminants is required, this may be the only option.

Notes to Appendix Seven

1. A. Turk, H. Mark and S. Mehiman, "Tracer Gas Nondestructive Testing of Activated Carbon Cells," Mat, Res, and Stds. Vol 9: pp. 24-26 (1969).
2. Air Pollution Third Edition, Volume IV – Engineering Control of Air Pollution, Are. Stem, Ed., Academic Press, New York, 1977, pp. 329363.
3. T. Godish, Indoor Air Pollution Control, Lewis Publishers, Inc., Chelsea, Michigan, 1989, pp. 293-298 & 301-303.
4. G.O. Nelson and A.N. Correia, "Respirator Cartridge Efficiency Studies: VIII. Summary and Conclusions," American Industrial Hygiene Association Journal, 37: pp. 514-525 (1976).

Appendix Nine

Steady-State Model for the Control of Gaseous Contaminants

A simple steady-state expression for a one-compartment model is helpful in identifying the three methods of gaseous contaminant control employed for thermal and air quality control.[1] A mass-balance for the model shown in Figure A11-1.1 may be expressed as:

$$\Delta C = (N\text{-}E)/ V_o$$

Where:

$\Delta C = C_r - C_o$ = The difference between the uniformly mixed indoor air concentration, C_r, and the outdoor concentration, C_o.

N = Q - S = The net generation rate of the contaminant where Q is the source strength (i.e., gross generation or emission rate), and S is the sink strength (i.e., sorption within the controlled space).

$E = {}_eV_mC_u$ = The removal rate of a contaminant in the air cleaner:

e = The removal efficiency (fractional) of the air cleaner in terms of the contaminant removed.

V_m = Volumetric flow rate of recirculated air.

C_u = Contaminant concentration upstream of the air cleaner.

C_d = Contaminant concentration downstream of the air cleaner (See Figure A11-1.1).

V_o = Volumetric flow rate of outdoor air for dilution control.

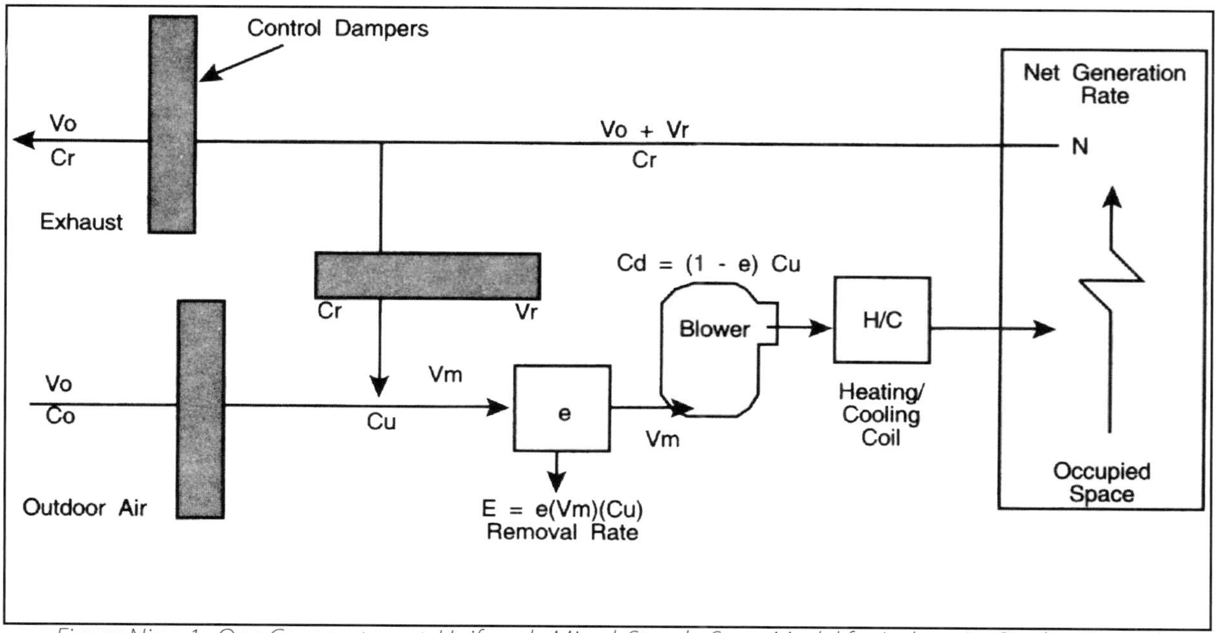

Figure Nine-1. One Compartment, Uniformly Mixed, Steady-State Model for Indoor Air Quality Control

In the model, the dilution rate, V_o, represents infiltration, natural ventilation, or mechanical ventilation with outdoor air. The removal rate, E, represents filter modules commercially available, or filtered 100% recirculated air commonly used in forced systems. In Figure AII1.1, V_r represents the volumetric flow rate of indoor air for dilution control.

Although this equation was derived from a simple model, it serves to identify some control strategies and their limitations for (indoor) air quality control:

- If *removal control* is not employed, the indoor (air) contaminant concentration will exceed outdoor concentration unless the *source* is removed or an infinite dilution rate is provided.
- If the contaminant concentration is to be less indoors than outdoors and the *dilution rate* is finite, the *removal rate*, E, must exceed the net *generation rate*, N.
- Outdoor air required for ***dilution control*** may be reduced, if alternative ***source control*** and removal control strategies are sufficient to provide the same quality of indoor air as would be achieved by dilution control.
- To achieve an acceptable ΔC, economically, a combined strategy of *source, dilution*, and *removal control* may be required.

Whether an indoor air quality control strategy employs an open- or closed-loop system, this equation indicates that the contaminant concentrations, C_r and C_o, must be known or specified, and that the net generation rate, N, must be known or estimated if indoor air quality is to be controlled directly. In this regard, it should be noted that Equation (1) is analogous to an energy balance in which N is compared to the total cooling load with an occupied space; ΔC to the change in enthalpy required to cool the supply air; and V_o to the volumetric supply rate of the supply air.

The above represents a very simple one-compartment model. However, even this simple model requires that both the ***indoor and outdoor contaminant concentrations*** and the ***net generation rate from within the space*** must be known or estimated to adequately control the indoor air quality. These three variables are the most important determinations as to what control strategy is to be employed.

Note to Appendix Nine

1. R.G. Matthey, T.K. Faison, and S. Silberstein, Air Quality Criteria for Storage of Paper Based Archival Records NBSIR 83-2795, U.S. Department of Commerce, National Bureau of Standards, Washington, D.C., 1983, pp. 22-24.

16

Glossary

A

Absorption: The process by which a liquid or a gas is taken into the interstices of a porous substance and held there, i.e., taken into pores of activated carbon where it becomes a liquid.

Acid: One of a large class of chemical substances whose water solutions may, among other things, turn litmus dye red, react with and dissolve certain metals to form salts, and have a pH of less than 7.0.

ACT: Alternative Control Techniques (EPA).

Activated alumina: Form of aluminum oxide used as a desiccant (dryer) for gases. It is also used as a carrier for potassium permanganate when the latter is used as a gas chemisorber.

Activated carbon: Form of carbon capable of removing certain gases from the air usually produced from exposing coal, coconut shell or wood to high temperature steam.

Activated charcoal: *See Activated carbon.*

Adsorbate: The gas which is removed from the airstream by contact with the adsorbent.

Adsorbent: An adsorber. That upon which adsorption takes place. It is the material to which a gas molecule is attached and retained.

Adsorption: The process by which gas molecules adhere to surfaces. The strength of the bond depends on the van der Waal forces between the gas and the surface.

Adsorption isotherm: A measurement of adsorption determined at a constant temperature by varying the amount of adsorbent used or the concentration of the adsorbate in contact with the carbon.

Aerosol: An assemblage of small particles, solid or liquid, suspended in air. The diameter of the particles may vary from 100 microns down to 0.01 microns. Examples: dust, smoke, fog.

Aerosol penetrometer: The complete assembly, including aerosol photometer, used to measure the penetration of aerosols through filter media. Also known as a Q107.

Aerosol photometer: A light-scattering mass concentration indicator.

Aerosol spectrometer: A device for measuring particle size distribution in air.

Agglomeration: The formation of a larger airborne particle by the collision of two or more smaller particles. Agglomeration takes place when the attractive forces between the particles is greater than the kinetic energy of collision.

AHAM: Association of Home Appliance Manufacturers.

Air: The mixture of gases that make up the atmosphere. Air is composed of 78% nitrogen and 21% oxygen. The balance consists of smaller amounts of gases which vary with the location in which the air is sampled.

Air change: A measure of the amount of air moving into and out of a space because of leakage or mechanical ventilation. One air change is a volumetric flow of air equal to the cubic content of the space. Example: If a space has a cubical content of 10,000 cubic feet and the ventilation rate is 1000 cfm , 0.1 (1000/10,000) air change is occurring every minute, or 6 (60 0.1) air changes are occurring per hour.

Air cleaner: A device used for the removal of particulate or gaseous impurities from the air.

Air conditioning: The process in which the physical condition of air within a space is maintained in a desired condition by controlling simultaneously its temperature, humidity, cleanliness and motion.

Air diffuser: An air distribution outlet designed to direct airflow into desired patterns.

Air dry: Type of curing process in which ambient air is used to carry off solvent vapor from wet paint.

Air filter: A device for removing particulate material from an airstream.

Air flow: Quantity of air (cfm) passing through a given cross sectional area (ft^2) at a stated velocity (fpm).

Air, makeup: Air supplied to a paint booth or building space for the purpose of replacing exhaust air or pressurizing that space with filtered replacement air.

Alkali: Any substance which in a water solution is bitter, is irritating or caustic to the skin and mucous membranes, turns litmus blue, and has a pH value greater than 7.0.

Ambient air: Air that surrounds the occupant or process in a space.

Apparent density: Weight of activated carbon per unit volume. Example: pounds per cubic foot.

Arrestance: A measure of the ability of an air filtration device to remove a synthetic dust from the air. ASHRAE arrestance is a measure of the ability of a device to remove ASHRAE test dust from test air.

ASHRAE: American Society of Heating, Refrigerating, and Air-Conditioning Engineers.

ASHRAE Synthetic Test Dust: A test dust composed of (by weight) 72% Standardized Air Cleaner Test Dust, fine (aka, Arizona Road Dust) predominately silica with a mean-mass particle diameter of 7.7 micrometers; 23% powdered carbon; 5% cotton linters #7 ground in a Wiley Mill with a 4 mm screen.

Atmospheric pressure: The pressure exerted upon the earth's surface by the weight of the atmosphere above it.

B

BACT: Best Available Control Techniques (EPA).

Bacteria: Single-celled microorganisms ranging from harmless and beneficial to intensely virulent and lethal.

Baffle: Plate or vane used to direct or control movement of fluid or air within a confined area.

Base coat: The color coat of material applied to a surface prior to the finish or clear coat.

Bed depth: The amount of adsorbent, expressed in length units, which is parallel to the flow of the airstream and through which the air stream must pass. Example: A tray type of adsorber may have a bed depth of 1 inch (25.4 mm).

B.E.T.: A method developed for measuring the surface area of granular activated carbon. It is named after the developers of the process: Braunauer, Emmett, and Teller.

Blow-through: An HVAC system with the fan located upstream of coils, air washer and final filters.

Blower: A fan used to move air under pressure.

Booth turbulence: The disruption of laminar flow due to imbalance or objects within the booth that deflect airflow.

BRI: Building-related illness. Illness whose cause is related to conditions inside a building.

Building related illness: Specific diagnosable symptoms such as skin rash or meningitis.

Brownian motion: The continuous zigzag motion of particles (aerosols) in suspension. The motion is caused by the impact of the molecules of the fluid (air) upon the particles.

Butane Activity: The maximum percentage increase in weight of a bed of activated carbon after air saturated with butane is passed through at a given temperature. The method is described in ASTM D5742-95.

C

CAAA: Clean Air Acts Amendments (1990),

CADR: Clean air delivery rate. ANSI (AHAM AC-1-2006).

Carbon dioxide (CO2): A gas consisting of one atom of carbon and two of oxygen. It is a product of oxidation of organic compounds. It is the principal metabolic product contained in the air we exhale.

Carbon filter: Air purifier using activated carbon as the air cleansing agent for the removal of molecular contaminants.

Carbon tetrachloride activity: The maximum percentage increase in weight of a bed of activated carbon after air saturated with carbon tetrachloride is passed through it at a given temperature. This procedure has been replaced by the butane activity test procedure, but is still referenced in some documents.

Catalyst: A material that promotes the rate of a chemical reaction and which may be involved in the reaction itself. However, at the end of the reaction it is essentially unchanged.

CCPG: Coatings and Consumer Products Group (EPA).

Central fan system: A mechanical system for heating, ventilating, or air conditioning in which air is treated or handled by equipment outside the space served and is conveyed to and from the space by means of a fan and distributing ductwork.

CFM: Cubic feet per minute.

CFU: Colony forming unit. A single microorganism or a cluster of microorganisms which when cultured on suitable nutrient will form a single visible colony.

Chemisorption: Removal of gases from an airstream by the chemical reaction of the gas with an impregnant on the surface of or distributed throughout the adsorbent or carrier.

Cleanroom: Room in which the concentration of airborne particles is controlled, and which is constructed and used in a manner to minimize the introduction, generation, and retention of particles inside the room, and in which other relevant parameters, e.g. temperature, humidity, and pressure, are controlled as necessary.

Coil: Any heating or cooling element made of finned pipe or tubing. The pipe or tubing may be connected in series or it may be installed in parallel between two headers.

Compressed air dryer: A machine or system used to remove water vapor from compressed air either through desiccation or refrigeration.

Condensation: Liquid or droplets that form when a gas or vapor is cooled below its dew point.

Contaminant: An unwanted airborne constituent that may reduce acceptability of the air.

Convection: Transfer of heat by means of movement or flow of liquid or gas.

Cooling coil: A heat transfer device that absorbs heat.

Cooling tower: A device that cools water by evaporating a portion of it in the air. Water is cooled to the wet bulb temperature of the air.

Crossdraft: The horizontal flow of air in a painting operation.

CTG: Control Technology Guidelines Curing oven: The chamber or enclosure in which heat is applied or created to fully cure the coating material.

D

Damper: A device for controlling airflow, usually in a duct.

Dehumidify: To remove water vapor from an airstream or from air in a space.

Delta P: *See pressure drop*. Delta (Δ) is the Greek symbol for change.

Denier: The weight in milligrams of 450 meters of fiber. Within fibers of the same specific gravity, denier can indicate fiber size. For example: a 12 denier round polyester fiber is 0.001328 in. (36 microns) in diameter. The round diameter in micrometers for fibers of the same density can be calculated by: DiA (u) = 11.88 x 59 Rt (Denier/ Specific Gravity). Example 12 Denier . (1.34 5.9)

Desiccant:: Media used in air dryers to remove water vapor from compressed air. Most commonly used media are activated alumina and silica.

Desorption: The opposite of adsorption. A phenomenon where an adsorbed molecule leaves the surface of the adsorbent.

Differential pressure: *See Pressure drop.*

Diffusers: Terminal devices that distribute conditioned air throughout a space. Diffusers can be aspirating (mix air from the space with the conditioned air) or non-aspirating.

Diffusion: The dispersal of air in an evenly distributed manner.

Diffusion Media: Filter media that by its density and resistance to air flow creates evenly distributed airflow (commonly referred to as laminar flow) through it.

DIN: The German Institute for Standardization (Deutsches Institut fur Normung e.V.). This institute establishes standards for testing and classifying air filters.

DOP: Dioctlyphthalate (diethylhexylphosphate), an oily liquid used in an aerosol form as a challenge for efficiency and leak testing HEPA filters. *See PAO.*

Downdraft: The typically ceiling to floor vertical flow of air in a painting operation.

Downstream: Area immediately after the filter.

Dry-scrubbing: The process of removing heavy concentrations of gaseous contaminants from an airstream by the use of specially designed adsorbers or chemisorbers.

Draw-through: An HVAC system with the fan located downstream of coils, air washer and filters.

Duct: Round or rectangular conduit through which air is carried from a central air conditioning system to various spaces in a building.

Dust: An aerosol of particles of any solid material, usually with particle size less than 100 microns.

Dust holding capacity: Amount of dust retained by a filter to its final recommended pressure drop.

E

Efficiency: (1) The ability of a device to remove particulate or gaseous material from an airstream by measuring the concentration of the material upstream and downstream of the device.

Electret: Filter media to which an electrostatic charge is applied during its formation.

Electronic air cleaners (two stage): Two-stage electrically powered filters. In the first stage the particles are charged and in the second stage they are captured.

Electrostatic precipitators: The type of air cleaners which gives particles of dust a charge by passing the dust-laden air through a strong (50- to 100-kV) electrostatic field. This causes particles to be attracted to oppositely charged plates so that they can be removed from an airstream. These devices are primarily used for stack gas cleaning.

Electrostatic filters (passive): A mechanical filter whose collection efficiency is augmented by the development of an electrostatic charge on the media by other than a continuous external power source. The electrostatic charge may be imposed at the time the media is manufactured (electret) or it may be generated by the flow of dry air through the media.

EPA: Environmental Protection Agency. This is the Federal agency that is responsible for setting standards for waterborne and airborne sources of pollution in the United States.

Evaporation: The process by which a liquid is changed to a gas. Heat is absorbed in the process.

Exhaust Filtration: Source capture media and/or equipment that prevents or limits emission of coating materials and/or volatile organic compounds (VOC).

Exhaust Velocity: Air flow, usually in fpm, at which overspray emissions and VOCs are removed from a paint booth.

Exfiltration: The controlled or uncontrolled leakage of air from a conditioned space.

Exhaust air: Air removed from a space and not reused therein.

F

Face area: The area of an air filter or other air treatment device normal to the flow of air through it.

Face Velocity: Average velocity of air through the face of a filter.

Fan: A radial or axial flow device used for moving or producing artificial currents of air.

Fan-coil: A terminal unit consisting of a finned tube coil and a fan in a single enclosure. These units may be designed for heating, cooling, or a combination of the two.

Fiber-free: (lint-free) The ability of a cloth product or filter media to retain its fiber integrity because of its construction.

Filter, air: A device for removing airborne contaminants.

Filter class: Filters classified as to their performance - MERV 1 - 16 according to ASHRAE 52.2, 6 - F according to EN 779 and H through U according to 1822.

Filter media: Material that makes up the filter element. Glass fibers and polyester fibers are examples of filter media. ("Media" is the plural of "medium." Common practice allows it to be used in the singular and "medias" as the plural).

Final pressure drop: Pressure to which a filter may be operated according to manufacturer.

Fog: An aerosol of fine water droplets in a gas.

FPM: Feet per minute abbreviated by the letter V representing velocity.

Fume: An aerosol of fine particles formed by the vapors of solid materials.

G

Gas: A fluid which has no fixed dimensions and fully occupies the space which contains it.
(2) Vapor phase or state of a substance.

Gasket (filter): Material used to prevent air leakage between filter media surface and its holding device.

H

HAP: Hazardous air pollution.

Heating coil: A heat transfer device which releases heat.

HEPA: High Efficiency Particulate Air (filter). A filter having a minimum efficiency of 99.97% on 0.3 μm particles.

Hogs Hair: Standard usage term used to refer to latex coated media composed of natural and synthetic fibers.

Housing: Device used to hold filter.

Humidifier: Device used to add moisture (water vapor) to air either in a space or in an HVAC system.

Hygrometer: A device for measuring the relative humidity of the air.

I

I-P: The inch-pound system of measurement, sometimes referred to as the English system of measurement, primarily used in the United States.

IAQ: Indoor air quality.

IEST: Institute of Environmental Sciences and Technology. A professional organization dedicated to the promotion of technology in several different areas. The Contamination Control Division of the Association has written several Recommended Practices for HEPA and associated filters and cleanroom equipment.

Impingement: The process in which particles are removed from an airstream because of their inertia. As air containing a particle flows toward a filter fiber or other collecting surface, the particle does not follow the air streamlines because of its inertia. Instead it moves in a straight line colliding with the filter fiber or surface to which it may become attached.

Inches of water gauge (in. w.g.): A unit used in measuring pressures. The equivalent measurement in SI is Pascals. 1 in. w.g. equals 248.8 Pascals (Pa).

Inertia: The tendency of a body in motion to move in the same straight line unless acted upon by some external force.

Infiltration: The controlled or uncontrolled leakage of air into a conditioned space.

Inorganic: Molecules whose chemical structure do not contain carbon. Not organic.

Interception: The process in which a particle is removed from an airstream as it follows the streamlines around a fiber. The particle comes in contact with a fiber and stays attached to it because the attractive forces between the fiber and the particle are stronger than the forces of disruption of the moving airstream.

ISO: International Organization for Standardization - worldwide federation of material bodies working through technical committees to prepare International Standards.

L

Laminar flow: Uniform, non-turbulent airflow moving from ceiling to floor or side to side in a clean room, isolation room or in a coating operation.

Laser particle counter: A device for measuring the size and quantity of aerosols in the airstream. It depends on the measurement of the amount of light reflected by individual particles.

Laser Spectrometer: *See Laser particle counter.*

Liter: A measurement of volume. A liter is 1/1000 cubic meter. 1 cubic foot equals 28.32 liters. The abbreviation for liter is "L."

M

Macropores: The pores in activated carbon that have diameters exceeding 1000 Angstrom units. An Angstrom unit is 1 10-8 meters.

MACT: Maximum Achievable Control Technology. Highest level of particulate removal through a contamination control system.

Magnehelic: Registered trade name for a diaphragm-activated dial and pointer gauge for measuring resistance to airflow. *See Manometer.*

Make-up air: Air that is supplied to a building to replace the air that has been removed by an exhaust system.

Manometer: Instrument for measuring pressure of gases and vapors. Gas pressure is balanced against a column of liquid in a U-shaped tube.

Media area: Gross: The total area of media used in the production of a filter. Net effective: The measure of usable media in a filter.

Media velocity: Volume flow rate divided by the media area fo the filter.

Median efficiency: In a series of efficiency tests, the median efficiency is the one which has an equal number of test results higher and lower than it.

Membrane filter: A solid continuous material (film) in which microscopic pores of controlled size are created by a variety of methods.

Metre (meter): An SI linear measurement equal to 39.37 inches. Metre is the preferred spelling.

Microbe: A microscopic single-cell organism.

Micron: One millionth of a meter or 1/25,400 of an inch. A micron is more correctly known as a micrometre (m).

Micropores: Pores in activated carbon ranging in diameter from 10 to 1000 Angstrom units. An Angstrom unit is 1 ten-billionth of a metre.

Mist: An aerosol formed by the dispersion of a liquid into very fine particles.

Modacrylic (fiber): A manufactured fiber in which the fiber-forming substance is any long-chain synthetic polymer composed of less than 85% but at least 35% by weight of acrylonitrile units [–CH2CH(CN)–]. Other chemicals such as vinyl chloride are incorporated as modifiers.

Mold: A fungus which grows on damp decaying organic matter. It is characterized by a fuzzy mat surface.

Molecular weight: The sum of the individual atomic weights of the atoms which make up a molecule.

Molecule: The smallest portion of an element or compound which retains the identity and characteristics of the element or compound.

Most penetrating particle size (MPPS): That particle size for which penetration through a filter is at its maximum. For smaller particles, the penetration decrease due to diffusion collection principles and for larger particles penetration decreases due to impingement and interception. MPPS is a function of filter media structure, air velocity and composition of the particle. It is the worst-case particle size with respect to filtration efficiency.

N

NAAQS: National Ambient Primary Air Quality Standards established by U.S. EPA.

Negative flow booth: A type of paint booth that operates with an exhaust fan only as a method of VOC/overspray removal.

NESHAP: National Emission Standards for Hazardous Air Pollutants. Regulatory standards based on national averages of pollution emissions for the control or elimination of pollutants through source capture or some alternative means.

NSF: National Sanitation Foundation. An organization that establishes standards for biological safety equipment and certifies technicians trained to maintain this equipment.

Nonpolar gases: Gases with symmetrical charges. No part of the molecule has a higher density of electrons than the others.

O

Odor: A quality of gases, liquids, or particles that stimulates the olfactory organ.

Odorant: A substance added to an otherwise odorless, colorless, and tasteless gas to give warning of gas leakage and to aid in leak detection.

Order of magnitude: The difference in two values measured by their logarithms. The quantity 100 (log 2.0) is an order of magnitude larger than the quantity 10 (log 1.0).

Organic: Molecules whose characteristics depend on the presence of one or more carbon atoms.

Optical particle counter (OPC): Term for white light or laser particle center.

Outside air: Atmospheric air; the air exterior to a refrigerated or conditioned space.

Overspray: Coating that does not contact the object being spray-coated and remains airborne as a particulate discharge.

"Owl": An optical device which measures the average particle size of an essentially monodispersed aerosol by the angular degree of light polarization.

Ozone: A gas whose molecules are composed of three oxygen atoms. Its symbol is 0^{3}. It is an unstable gas that is significantly toxic. The 1989 threshold level value for ozone was 0.1 part per million for an 8-hour time weighted average.

P

PAO (Polyalpha Olaphins): Challenge liquid used to replace DOP.

Particle bounce: Describes particles that impinge on a filter fiber but are not retained.

Particle count: Number of particles of known sizes in a given volume of air, usually measured with a particle counter.

Particle counter: A device for measuring the number and size distribution of particles in a fluid, usually abbreviated OPC for optical particle counter.

Paint arrestor: A filtration system used to trap and hold overspray in a coating operation. It may be a water cascade type, a mechanical (media or baffle) type or a recovery system.

Paint Holding Capacity: The total accumulation and retention of coating materials on a paint arrestor when it has reached a stated pressure drop. It may be defined on a wet or dry basis.

Paint Removal Efficiency: The percentage of total paint or coating material stopped or trapped by the arrestor versus the total challenge volume.

Pascal: The SI standard for measuring low gas pressures. By definition it is N/m2 (N Newtons, m2 square meters). 1 in. w.g. 248.8 Pa.

Passive electrostatic filter: *See Electrostatic filter (passive).*

Penetration: A measure, in percent, of the material passing through a filter. Mathematically penetration is 100 - Efficiency (percent). If a filter is 98% efficient, its penetration is 2% (100 - 98). Penetration is used to measure the performance of very high efficiency filters.

Photometer: A device that measures the mass concentration of an aerosol by the amount of light the aerosol scatters.

Pitot Tube: A device used to measure the velocity pressure of an airstream by simultaneously measuring its static and total pressures. Velocity pressure is the total pressure minus the static pressure. Velocity of air at standard conditions can be calculated by using the formula:

$$V = 4005\sqrt{VP}$$

Where:
V = velocity of air (fpm)
VP = velocity pressure (in. w.g.) from pitot tube reading.

Plenum chamber: An air compartment maintained under positive or negative pressure and connected to one or more distributing ducts.

PM: Particulate matter.

Pocket filter: name used for bag filter.

Polar gases: Gases with asymmetrical charges. One part of the molecule has more electrons than the rest.

Potassium chloride (KCl): Challenge aerosol used in ASHRAE 5.2.2 particle size efficeincy test.

Potassium permanganate (KMn04): An oxidizing agent impregnated on activated alumina and other inert carriers. *See Chemisorption.*

PPI: Pores per inch. A measurement of the porosity of foam materials.

PPM: Parts per million.

Pressure drop: The resistance of a device to the flow of a fluid through it. The pressure drop of a filter is a measure of its resistance to airflow through it. Resistance is measured in inches water guage (w.g.) in the Inch-Pound system of measurement. It is measured in Pascals in the SI system.

PSL: A suspension of polystyrene latex spheres all of which are uniform in size. They are used to calibrate some test instruments or used as a challenge agent in some filter tests. They are available in a variety of micro-sizes.

R

RACT: Reasonably Achievable Control Technology. A standard by which the discharge of airborne pollutants is controlled with "reasonable" efficiency, by filtration systems or other methods.

Rated filter capacity: The specific quantity of air recommended by a filter manufacturer to be handled by a filter.

Reactivation: The removal of adsorbates from spent granular activated carbon, allowing the carbon to be reused. This is also called regeneration.

Recirculated air: Air that has been taken from the space, reconditioned (temperature, humidity, and cleanliness adjusted as necessary), and returned to the space.

Refrigerated dryer: A unit that condenses water from compressed air through a process of refrigeration and cooling.

Resistance: *See Pressure drop.*

RP: (1) Recommended Practice. Recommended practices are equivalent in content to Standards but do not carry the strict requirements of compliance that Standards do. (2) ASHRAE research proposal.

S

SBS: Sick building syndrome. Term applied for building where occupants have non-specific symptoms, i.e., flu-like symptoms.

SI: International System of Units. In HVAC the more common units are: Litre (L) - volume; second (s) - time; metre (m) - distance (A kilometre is 1000 metres. A millimetre is 1/1000 metre); degrees Celsius (°C) - temperature.

Smog: A mixture of gases and aerosols generated from a variety of sources. Smogs have a damaging effect on respiratory organs of people. Most smog is attributed to the exhaust from automobiles and to the discharge of industrial processes.

Smoke: (1) An aerosol of particles usually but not necessarily solid, formed from combustion or sublimation. (2) Carbon or soot particles less than 0.1 micron in size which result from incomplete combustion of carbonaceous materials such as coal, oil, and tobacco.

Standard velocity: The rate of airflow in a surface coating operation as dictated by OSHA and/or design parameters. It is usually 50-175 fpm.

Static pressure: The potential pressure exerted in all directions by a fluid. For a fluid in motion it is measured in the direction normal to the direction of flow. It has the potential to either burst or collapse a duct or other enclosure.

Static tip: Device inserted at right angles to an airstream to measure static pressure.

Stratification of air: Condition in which there is little or no air movement in a room; air lies in temperature layers.

Sublimation: The conversion of a solid phase of a substance to a gaseous phase without going through the liquid phase.

SULPA: Super Ultra Low Penetration Air (filter). These filters typically have efficiencies of 99.9999% on its most penetrating particle size. Since penetration measurements at this sensitivity is beyond the capability of D.O.P. testers, other methods must be used.

Supply air: A mixture of recirculated air and outside air which has been conditioned and delivered to the space. Supply air can be 100% outside air or 100% recirculated air.

Surface area (carbon): The surface area of granular activated carbon is determined by the BET method which utilizes the adsorption of nitrogen at liquid nitrogen temperatures in the calculation. Surface area is usually expressed in square meters per gram of carbon.

T

Tackifier: A substance applied to filter media to increase the retention of dust. It can be applied to the surface of media or throughout its depth. It may be an oil, a pressure-sensitive resin, or a solvent which imparts a tacky surface to the media.

Total pressure: Also called impact pressure. The pressure measured in a moving fluid by an impact tube. It is the sum of the velocity pressure and the static pressure.

Transfer Efficiency: A measure of the amount of coating material that actually covers the product versus the total amount used. It is expressed as a percent.

U

ULPA: Ultra Low Penetration Air (filter). Filters in this category typically have efficiencies of 99.999% on 0.3 micron DOP particles. They may be rated and tested for penetration on other size particles.

Upstream: Region where airflow is approaching the filter.

V

Valence: The number of electrons which an atom can give or can take up when reacting with other atoms to form a chemical compound.

van der Waals forces: The forces of attraction between molecules.

Vapors: Gases formed by the evaporation of materials which are normally liquids or the sublimation of materials which are normally solids.

Velocity: The distance traveled in a given time. Air velocity is measured in feet per minute (fpm) or meters per second (m/s).

Velocity pressure: The kinetic pressure in the direction of flow necessary to cause a fluid at rest to flow at a given velocity. *See also "Pitot tube."*

Ventilation: The introduction of outdoor air into a building by mechanical means.

Virus: A microscopic particle composed of DNA or RNA surrounded by a protein coat. Viruses replicate by entering a cell, releasing their own DNA or RNA, and controlling the DNA of the cell to manufacture more of the virus DNA or RNA and the surrounding protein coating. In the process the cell is usually destroyed, its walls disrupted, and the virus released into the surrounding environment.

VOC: Volatile organic compound. An organic compound which evaporates at room temperature.

W

"W.G.": *See Inches water gauge.*

Wet wash: A paint collection device utilizing a vertical cascade of treated water as a filter medium to capture paint overspray.

Z

Zeolite: Type of silicate adsorbent used for removal of certain odors. It is also used as a carrier for such reactive chemicals as potassium permanganate.